Shirts of Steel

SHIRTS OF STEEL

An Anatomy of the Turkish Armed Forces

MEHMET ALI BIRAND

With a Foreword by
William Hale

Translated from the Turkish by
Saliha Paker and Ruth Christie

I.B. Tauris & Co Ltd
Publishers
London · New York

Published in 1991 by
I.B. Tauris & Co Ltd
110 Gloucester Avenue
London NW1 8JA

175 Fifth Avenue
New York
NY 10010

In the United States of America
and Canada distributed by
St Martin's Press
175 Fifth Avenue
New York
NY 10010

A CIP record for this book is available from the British Library

Library of Congress Catalog card number: 90–071578
A full CIP record is available from the Library of Congress

ISBN 1–85043–326–7

Printed and bound in Great Britain by
Butler & Tanner Ltd, Frome and London

Contents

Foreword

Over the past twenty years or so, a number of books have appeared attempting to explain the political role of the military in Middle Eastern as well as African, Asian and Latin American countries. They have given us invaluable information, as well as fascinating perspectives on a social and political institution which still dominates the politics of a large number of Third World states. However, they are mostly the work of scholars in North American or European universities who are not direct participants in the societies they are describing, addressing a limited and mostly academic audience, and sometimes over-burdened with elaborate conceptual frameworks.

Mehmet Ali Birand's book is different. The author is one of Turkey's most influential and respected political journalists. He has previous books on the Cyprus problem and Turkey's relations with the European Community to his credit, as well as a fascinating study of the events leading to the military coup of 12 September 1980 which has also been published in English (as *The Generals' Coup in Turkey: An Inside Story of 12 September 1980*, London, Brassey's Defence Publishers, 1987). In Turkey, he is best known for his work as chief European correspondent for the popular liberal daily *Milliyet*, for which he also writes a regular column on foreign affairs, and as presenter of the monthly TV current affairs programme *Thirty-second Day*. In its original Turkish edition, which was published in 1986, this book went through no less than 15 reprints, selling a total of 75,000 copies.

The reason is not hard to find. Although virtually all adult male Turks go through a period of eighteen months of military service, and the Armed Forces have played a central part in the nation's political life, critical analysis of their performance and functions has been regarded as taboo. By breaking this convention, Mehmet Ali Birand excited the interest of a very wide audience. Moreover, by allowing the officers whom he interviewed to speak for themselves, and by following through the career of a typical officer, from his initiation into the army at the age of 14 to his final retirement, he gives his account a liveliness and directness which is lacking in most academic studies. *Shirts of Steel* gives insights into a hidden world, as unknown to most Turks as it is to foreigners. Turkey's history, and its strategic position in the borderland between Europe and the Middle East, heightens the importance of his study. For those concerned with analysing the role of the military in politics in other parts of the world, the Turkish story is also an important test case.

As the book makes clear, the Turkish Army's historic role in the Ottoman Empire plays an important part in the education programme for fledgeling Turkish officers. In more recent years, its position as a political actor has been shaped by the three incursions into the political system which it made in 1960, 1971 and 1980. Knowledge of these events is second nature to most Turks, so they are referred to only in passing in this book. However, they will be less familiar to most foreign readers. Hence, a brief historical summary seems in order.

In the Ottoman Empire there was an almost complete identification of the army with the state, whose servants were in many cases both soldiers and administrators. The Janissaries, the trained regiments of professional infantry who formed the core of the army, were officially 'slaves of the Sultan'. They were originally recruited from the Christian minorities by a system of conscription known as *devshirme*, in which young men were taken from their distant homes in the Balkans, and inducted into a new world – a Muslim, military world – in which the Ottoman state became their family as well as their master. The process has striking parallels with the modern pattern, in which the Turkish Army still draws in a substantial proportion of its future officers at a tender age, and provides them not just with a profession, but also a separate social milieu which lasts for the rest of their lives. From the Ottoman Empire's now distant past, they inherit an elite status, but they are now faced with the task of reconciling it with a modern industrializing society which has thrown up new elites and new challenges. Cast aside from the social as well as professional nexus of the army, officers are like fish out of water, as Birand's final chapter poignantly illustrates.

During the nineteenth century, the army was the first arm of the state to be affected by the programme of modernization, by which the later Sultans sought to stave off the challenge of the advancing European powers. Hence, the officers of a new army were projected into the vanguard of the campaign for change, playing a crucial part in the introduction of Turkey's first constitution in 1876, as well as the Young Turk revolution of 1908. However, the second of these was to lead to disaster. After 1908, the Young Turks became became hopelessly divided, and both the army and state structure collapsed with them. The political partisanship which divided the officers corps and led to the humiliating defeats of the Balkan War of 1912–13 is a lesson which is still impressed on Turkish officer cadets, and formed the prelude to the greater tragedy of 1914–18.

The turning point came with the dramatic events which succeeded the Great War, in which a new Turkish state was built on the ruins of defeat. In Ataturk's vision, the new Turkey was to be a national republic, divorced from its Ottoman past, and basing its legitimacy on the concept of popular sovereignty in place of Islamic tradition. The defeat of the occupying entente powers in 1920–2 also gave the army an heroic stature as defender of the nation. Once the victory was secured, however, the Armed Forces were left in an ambiguous position. On the one hand, Ataturk was determined to keep the army out of the political system, to make sure that the army itself was not divided by politics, and that ambitious officers could not challenge his leadership. On the other hand, he continued to look to the army as the ultimate guardian of his achievements. As Mehmet Ali Birand makes clear, Ataturkism is instilled almost as a religion among army officers. However, its exact implications – whether, for instance, the concept of guardianship takes precedence over that of non-involvement in politics – raise questions which today's officers still find it hard to answer.

In the years after 1945, there was a transformation of Turkey's internal political regime, as Ismet Inonu, Ataturk's successor as President, oversaw the transition to a multi-party regime. The test of the army's adherence to the principles of political liberalism came in 1960 when a group of officers, convinced that Prime Minister Adnan Menderes was undermining Ataturkism, overthrew the government in Turkey's first post-war coup d'état. The military junta which ruled Turkey during 1960–1 was riven by a division between those officers – mostly at the upper end of the age/rank scale – who favoured a return to elected civilian government, and a group of younger radicals, who wished to set up a long-term authoritarian reformist regime. The victory of the first group, with the return to civilian politics in 1961, can be attributed to two of the factors in the army's structure and culture to which Birand draws attention, that is, its strong sense of hierarchical discipline, and the conviction, inherited from Ataturk, that long-term direct involvement in politics would have fatal effects on its homegeneity, by dividing the officers on partisan lines. This point was underlined by the abject failure of two attempted counter-coups by one of the radicals, Colonel Talat Aydemir, in February 1962 and May 1963. The fact that, on the second occasion, Aydemir had drawn virtually his sole support from the cadets of the Ankara Military Academy led to some major upheavals in the military training system, as Birand relates.

After 1963, the Generals and politicians settled back into a state of sometimes uneasy cohabitation which lasted until 1971. By this stage, a new group of reformist officers had become convinced that Suleyman Demirel's government was intentionally obstructing the implementation of social and economic reforms which had been outlined in the constitution of 1961, but had remained a dead letter. Other, more conservative Generals were alarmed by the government's failure to curb rising violence by urban guerrillas, mostly of the left, which undermined law and order after 1968. The two groups compromised by issuing a *pronunciamento* which forced Demirel's government to resign, putting in its place a supposedly supra-party administration, which followed the off-stage directions of the military chiefs. By 1973, however, the Generals had realized that this kind of quasi-military regime was an unworkable alternative to either outright military rule, or freely elected civilian government. Unwilling to take over power openly, they retired to their barracks in October 1973, to give way to a series of weak, but at least elected, coalition governments.

The problems which led to the third military intervention, that of 12 September 1980, were far more acute, and the legitimacy of the coup was far more widely accepted. By 1980, civilian government had almost completely broken down, as a renewed surge of politically inspired violence threatened to drag Turkey down into a Lebanese-style collapse of the state. During the first seven months of 1980, over 1,250 people were killed by politically inspired desperadoes of either right or left. The economy was in chaos, with treble-digit inflation, a huge foreign trade deficit, and a rampant black market in essential commodities. The civilian political leaders had failed to heed repeated warnings by the military to sink their mutual rivalries by forming a national unity government. As a result, when the army struck, the coup was greeted with general relief.

Under General Kenan Evren, the army established a five-man junta which restored law and order and successfully tackled the country's economic problems by bringing down inflation and the foreign trade deficit. Besides the thousands of terrorists who were rounded up, there were large-scale arrests and trials of those who had done no more than express oppositional views. Before returning power to the civilians, the military regime also drew up a new constitution which, by restricting civil liberties, aimed to install a more restricted version of democracy than had been sanctioned since 1961. This commitment reflected the tensions within the political culture of the military to which Birand refers – that is, the official endorsement of democracy, combined with the solidarist ideal of a disciplined, mutually cooperative society.

These developments left Turkey with a host of unanswered questions. Would the return to a civilian government under Prime Minister Turgut Ozal in 1983 be a genuine one, or would the military continue to control the country's political destinies from behind the scenes? Would civilian governments achieve proper control over the Armed Forces, as in the Western democracies, or would they continue to act as a semi-autonomous arm of the state? Above all would the ten year cycle of intervention, withdrawal, political breakdown and re-intervention, which had been repeated three times since 1960, be finally broken?

Tentative answers to some of these questions are suggested by changes which have occurred since 1983 – in many cases, since 1986, when this book was first published. During this time, the army has steadily withdrawn from the political scene. The military regime's attempt to encourage the establishment of a civilian government in its own image proved a failure, as the party supported by the military was roundly defeated in the 1983 elections. The pre-1980 political leaders, whom the army had tried to exclude from the system, regained their political rights in 1987, following a national referendum. Finally, in 1989, when ex-General Kenan Evren retired from the Presidency, he was duly succeeded by Turgut Ozal, Turkey's first fully civilian head of state since 1960. Most importantly, all of these changes were accomplished without any overt protest from the military – in fact, apparently with its full support. Whether this trend will be a permanent one is still far from certain. The outcome will probably depend as much on the actions of the politicians as on those of the Generals. It will also require far more openness in the relations between the two sides than has obtained in the past. By opening up the Turkish Army to the outside world, Mehmet Ali Birand's book marks an important advance in this process.

William Hale,
School of Oriental and African Studies, London.

Preface

The Turkish Armed Forces, 800,000-strong and absorbing 25 per cent of the national budget in 1986, constitute the best-organized, best-disciplined and longest surviving establishment in the country. They are, moreover, one of the largest armies in any NATO country.

The Armed Forces are a force that has influenced the Turkish people's political and daily life by a number of interventions, and they will make their weight felt in the future too. In Turkey, the question, 'What does the army say?', crops up constantly in private conversation and political argument, and requires a response. Episodes of political intervention and how they came about are a major topic in newspapers, magazines and books.

The Turkish people as a whole are proud of their army, want it to be strong, and accord it a status which no army enjoys in any other NATO country. In Turkey, the army is always praised, never criticized, and, in an emergency, it is seen as the nation's saviour. The truth is, however, that the Turkish public knows very little about the guiding force of this gigantic body – the 35,000-strong officer corps. In no other civilized country in the world is the army so little known and yet so close to the hearts of the people. There is a great deal of ignorance of the background of those who affect the daily lives and security of the Turkish people so closely, of their training, the ideas they grow up with, their routine, their priorities – in a nutshell, what an 'officer's world' is like. Nor do those who do their one-and-a-half years of conscript service, in the ranks or as reserve officers, have any adequate picture of this 'world'. My reason for writing this book was to attempt to shed some light on the subject and fill this gap in our knowledge.

I did not set out on the quest, which eventually took five years, with high hopes. To begin with I studied how Western armies trained their officers and then, over the last two years, I focused my attention on the Turkish Armed Forces. Close friends warned me that I was 'on a fool's errand', and that 'Turkey was not ready for this yet', stressing that the army was a taboo subject. So I put the matter to the direct test and wrote to the General Staff, requesting information and assistance for the study I had in mind. The answer confirmed my belief that the army did not want to be a taboo subject. It was the civilian population who had chosen the easy option of setting up the army as sacrosanct.

I began to gather the kind of information that had been dealt with in hundreds of books and articles in every civilized country except Turkey. I visited Military Academies, ate at their tables, had long discussions with the officers and military students. In short, I shared their life in the barracks, and that is how this book came to be written. You will see the Turkish Army through the eyes of the officers themselves. It was of course impossible to talk to everyone individually. What I did was to present the views that 'lent themselves best to generalization', and among them you may encounter a whole range of opinions, for the members of the forces are as different from each other as the fingers of your hand. It is, nevertheless, quite possible that I have occasionally gone too far in generalizing and have failed to present minority views sufficiently.

I have purposely left out the names of the people I interviewed, one reason being the promise I made to the individuals concerned and another the reluctance of officers to have their names bandied about 'as if they were seeking publicity'.

Once completed, the book came under no official scrutiny (nor was there any request to that end by the General Staff). But I did have the manuscript cross-read by certain intimate friends, both retired and still serving, whose views and opinions I value highly, 'to check the book for errors and go over its technical aspects'.

*   *   *

I wrote this book neither to praise nor to run down the Turkish Armed Forces. The army is essential to national security and it is obvious that unnecessary strictures will get us nowhere. My aim in writing it has been to make one of Turkey's most important establishments better known, and to draw attention to a number of questions each important enough to merit a whole book to itself.

The Turkish Army is part of the Turkish people and its mirror image. As a result, the contradictions, the maladies and the backwardness that afflict Turkish society afflict the army too. My aim has been to present the most accurate picture possible of the Turkish Armed Forces, the facts as distinct from the daily hymn of praise, and thus help the reader to arrive at a realistic assessment. It is quite possible that this book may lead you to think, 'I never knew the Turkish Army had so many problems'. That would be a misguided comment; all armies have similar problems. The only difference lies in whether these problems are open to discussion.

I owe a great debt of gratitude to a whole range of people who helped me with this book. They include those who trusted me with their personal views, from the students at the Military Academy to their commanders, from second-lieutenants to top generals, from retired officers to ex-Ministers of Defence. They also include officials of the US Defence Department, and of the British, Belgian, German and French Ministries of Defence, who let me carry out research in their own military schools, provided me with observations on their own armed forces and on the Turkish Army, and supported me with documents and research facilities. I am indebted to Hikmet Bila who gave unstinting support during the preparation of the book, to Ahmet Baydar whose untiring research enabled me to write certain sections, and to Fehiman Cebeci for his meticulous research.

Part One
THE MAKING OF AN OFFICER

Preamble
Birth of a Commander

You have entered a great academy of discipline: we should all be congratulated. From now on your lives will change . . . If you work according to the discipline, obeying orders and proving yourselves worthy, we will provide you with a profession superior to any other . . . one which cannot be acquired through money or possessions . . . You will be performing the most illustrious duties in the world . . .

The Commander of the Academy was speaking. There was utter silence. Everyone listened with concentrated attention. The successful candidates gathered in the conference hall were looking at this imposing personage whose position they had not yet quite understood. A row of flags was ranged behind him of which they recognized only the Turkish. The other officers, in immaculate uniforms, were standing at respectful attention behind the commander. He continued:

Your flag will be the great Ataturk. Your ideology will be his principles, your aim will be the direction he showed us. You will follow unswervingly in Ataturk's footsteps.

The new cadet looked around. He knew about Ataturk but not much about his principles and he wondered if they would be questioned immediately on the subject. From the expressions of those about him he concluded with relief that the majority did not know much either.

You will dedicate yourselves unconditionally to your country. You will give all your thoughts to your country alone and put yourselves and your family in second place. You will preserve our glorious flag over this land and give your lives to guard and protect this sacred earth . . . Ataturk put this country in our safe keeping. If you become a soldier and officer worthy of Him we will entrust the flag – which means the fatherland – to you. This country will belong to you.

His heart beat fast. Nobody had ever suggested that such a responsibility would be handed to someone like himself.

One day you will eat at the tables of kings along with the most distinguished company and another day you'll share your rations with an ordinary Turkish soldier. It is for you to set an example.

The prospect pleased him and he imagined himself alighting from large cars: 'If the family could see me how pleased they'd be', he thought.

From this day forward you have become a member of the Turkish army renowned for its historic deeds of heroism. Your uniform represents the army. One mark on your uniform is a mark of dishonour for the army. Every step requires care. By a mere word or action you can either glorify or disgrace your uniform and our famous army. It is your duty to glorify. Those who do so will remain with us to carry on this splendid task, but those who do not will find themselves out in the cold.

He felt himself trembling and wondered what would happen if he was thrown out. The first thought that came into his head was, 'I'd die'. The people of his neighbourhood and, more important, his father's face came before him. He fell into a cold sweat.

There is only one way for you to succeed; to strive with all your might and strength, to maintain discipline and obey orders. With us discipline is comparable to commitment to a religion. Disobedience and lack of discipline are the least forgivable faults.

You have seen during the examinations how many thousand men applied for places. No one must take up a place if he is not worthy.

Up to these final words everything had been fine; now he suddenly got the feeling that this business was not quite what he had expected. He felt very much alone and missed the affectionate warmth of his family.

'What am I doing here?' he thought and realized it was too late to wonder. He had really liked the Academy, which was a splendid building such as he had never seen before. Moreover, they had given him brand-new clothes. The halls were comfortably warm and incredibly clean.

From this moment I ask you all to become worthy of this Academy.

The Commander had finished his speech. The high-ranking officers about him saluted and once again the prospects were pleasing. The cadet said to himself, 'Please God help me to uphold the honour of my family', and with that he embarked on his first lesson, the first step into a world completely new to him and of which he had not even been aware.

1
The Attraction of a Military Career

'Why did you choose a military career?'

My question is addressed to a bright-eyed cadet, aged fifteen. He has been in Military School only a few months but he already knows the routine answer: 'Because I love it, Sir!' To him, everybody is still 'Sir!'. I smile at him. He smiles back.

'Let me put the question once more. What made you want to be a soldier?'

This time, he relaxes his guard a little. 'I don't know. My family were very keen. My dad said it was a very good profession, which could take me much further than I would ever reach if I stayed near home.'

'How about you? Are you happy?'

'Very.'

'And what do you like best about it all?'

'Going out in uniform.'

The military boarding schools constitute the basic source of the Turkish officer corps, and currently supply half the intake for the Military Academies, while the other half comes from graduates of civilian schools. This makes the Turkish system, initiated in 1845 under the Ottoman Empire, unique, with an unbroken run of over 140 years.

There is a marked difference between the youths who enter these colleges as cadets at the age of fourteen or fifteen and the young men who enter the Military Academies at nineteen or twenty. The former are not so much influenced by personal ambition as by their family, whereas the direct entrants to the Military Academies have a clearer motivation, though a large number of them too are influenced by their families.

When I came to Kuleli Military High School for the admission
tests, I had no idea where I was or why, or what the school was
like. My dad had led me there by the hand.

When I sat the entrance examination for the Naval School I
was not really aware of the implications. I made the ridiculous
discovery after passing the exam that I was afraid of the water,
that I wasn't a natural swimmer, and the slightest motion of the
sea made me sea-sick. But by then the die was cast and I couldn't
quit. In a military school you can't just give up. It was just as well,
though. I came to like my career very much.

The first of these speakers is one of the most outstanding Staff
colonels in the army and the second is a rear-admiral.

WHY A MILITARY CAREER IS HIGHLY REGARDED

A military career is not very highly regarded in the industrialized
Western countries, and is even unpopular. A military man,
particularly in Western Europe, is seen as someone who might
involve his country in war rather than function as its defender.
The Second World War, which resulted in the deaths of some
50 million people, has had an important influence on the
development of this anti-militarist attitude. Economic, cultural
and social factors, and particularly the way the new generation
has been brought up, have considerably reduced the status of
the 'soldier'.

This swing has been so strong that now in the West
(particularly in countries which maintain professional armies)
recruitment to the armed forces is becoming increasingly
difficult. In the light of the prevailing low birth-rate, a multitude
of incentives are being tried out to overcome the problem. The
old campaigns based on the theme of 'Your country needs you'
are now giving way to slogans like 'Join up now for good pay,
unequalled opportunities and the chance to visit foreign lands'.
More and more countries are attempting to recruit through the
press and even television.

In Turkey the situation is quite the reverse. In the period
1970–86, for instance, some 35,000 civilians applied for

admission to the Army Academy and only 5,100 were successful. Admittedly, a major sector of the Turkish people has a traditional feeling for a military uniform and a love of soldiering that borders on passion. Nevertheless even the most cursory glance reveals that these are not the only factors shaping the Turkish attitude towards a military career; there are more important influences at work.

> On my father's death, my mother was reduced to looking after her four children on a widows-and-orphans pension. My elder brothers went out to work but that wasn't enough to take care of me too. My sole opportunity for an education lay in a military school, as it was completely free.

This boy is now a military attaché.

> My father was a train conductor. He died when I was not quite two. My mother had no educational qualifications and my sisters had to abandon their education and go to work. I had just one thought in mind, not to be a burden to my family. We were on the borderline of poverty and I knew I'd be finished if I didn't get an education. The only solution was to go to the Military Academy.

This young man is on his way to becoming a general.

> I'm a teacher and know how important education is. To know a foreign language is worth several university degrees. There was no way I could see my son through university on my salary. If, however, he could enter a military school, everything would be free, board and lodgings and clothing. You don't even have to pay for textbooks. On top of that, he is even paid pocket-money, not much, but enough. It's true that now we enjoy him for only forty-five days in the year, but so what? One has to pay a price for everything.

The fact that the military schools provide a free education is, in the context of the economic realities of Turkey, one of the strongest motivations for the families concerned. The 'economic aspect' of the attraction of the military career becomes obvious if one takes into account the fact that 75 per cent of the families of the cadets have a monthly income of 30,000–95,000 Turkish lira, according to the latest statistics.

The high quality of instruction at the military schools is also an important consideration, though not to the same extent as the economic element. The teaching of foreign languages in particular, and the importance attached to this in the past few years, is a consideration which seriously influences the decisions taken by many families.

> Where we lived, we had a hard time scraping together the means to attend a high school, let alone the university. In addition to the cost, our child would have had to leave our small town to go to the city. And even then I would have no idea of the quality of the education he would be receiving. But I've heard that children at military colleges get a good education and proper nutrition – and even have running hot water.

According to the applicants, the inadequacy of the civilian schools and the superiority of the education available at military schools is an important and influential element in their choice.

That, of course, is not all. An additional factor emerged very clearly from the following statement by a tradesman who incidentally had fainted with excitement the day he succeeded in getting his son admitted to the Military Academy.

> I only hope he doesn't waste his time and completes his studies. He is not aware of it yet, but what a guaranteed future lies ahead of him. Unless he makes a fool of himself, he'll be receiving a regular income for life. I have officer friends and I know. While they might not be earning much, they have enough to meet their basic needs. I don't want my son to be in an occupation like mine where the future is uncertain.

While 'love of country, love of soldiering, and the special feeling of belonging to a military nation' do occupy the hearts of some of the families who want their sons to enter one of the Military Academies, for the vast majority the reasons are quite different. Lack of money, the lack of quality and equality of opportunity in education, and the desire for a secure income, have persuaded workers and retired workers, the self-employed, and also civil servants, especially teachers, to favour the army as the best career.

One should not immediately think of the self-employed as merchants or industrialists. Small shopkeepers and the heads of small family businesses are keen for their children to achieve a more secure way of earning a living, and small businessmen figure substantially among the self-employed.

Therein perhaps lies the root cause of the difference between the Turkish Army and the armies of many other countries (those of Latin America in particular). All the characteristics of the Turkish officer derive from middle-class influences, from his manner of speech to his world outlook, from his attitude in performing his duties to the way he regards civilians and politicians.

These, then, are the reasons for the choice of career by the 35,000 officers who form the backbone of the Turkish Army and who will command a gigantic body of 800,000 men. The following words by a commanding officer probably sum up the real situation:

> Young men do not come to soldiering because they are in love with the career but on the whole to find employment and security. We are here to teach them the profession and get them to love it. Then we come face to face with the problem prevalent in Turkey: the problem of resources and the problem of human resources. When, some day, Turkey's economy improves and unemployment falls, those who love the military profession for its own sake will begin to join us and we shall become more productive.

2
A Total Change From the Very First Day

Using the same clay, we craft a fine vase at the military schools. Civilian schools produce jugs of poor quality.

A commander's personal view

From the moment the classroom doors close behind him and the first lesson begins, the young cadet senses that his life will be transformed in a completely unexpected way. Immediately afterwards come a series of very striking 'firsts' for the majority of students. One of the earliest problems encountered in the Military Schools and Academies that admit civilian students is, in particular, the sense of loneliness and withdrawal that affects the young man who has left his small town, village, or neighbourhood.

The excitement of sitting the entrance examinations and the great pleasure for the candidate and his family when he passes are suddenly lost. Alone in those enormous corridors he feels depressed. Tears well up in his eyes, and he recalls his mother's affectionate embrace; but he buries his pain and withdraws into his shell.

When I went to bed at night I felt like crying out loud. I missed my mother terribly and I used to dream of my father, in spite of his strictness. I used to pray silently, asking God to re-unite us. It was the sort of exquisite pain that must be experienced to be understood. Particularly when, on my first night away from home, I discovered in my pocket a note my father must have slipped me in secret. It said, in his poor untutored hand, 'My

11

boy, do not disgrace our neighbourhood. Remember we love you and put yourself in God's care.'

That was the father who, when his son was admitted to military school and was returning home in uniform to make his farewells, hired a taxi and rode up and down the main street of the small town several times to show off his son. What a boost that had been to the father and, perhaps, an exhibition of superiority.

I kept all these feelings to myself. Anyway, I was afraid of going out into the city and I stayed in the school at weekends, nervous of getting lost. I almost had a feeling that this big city was going to crush me. This went on until one day my counselling tutor asked me to come and see him.

The busiest time for the counselling tutors is in the first months of the new school year, particularly with the new students. They take in hand these tender youths cut off from their native surroundings, put in a place they have never seen before, and try to acclimatize them until they can stand on their own feet.

The young entrant's idea of his future commander's severity (which I prefer to see as strength of perseverance) dates back to his earliest days and will never be lost. As he gets used to the regulated and disciplined conditions of the Military School, which are so incompatible with the feelings of family kindness, his life begins to change a little more. From the age of fourteen he is taught that crying is a great weakness and even if he is homesick he realizes he must not show his tears. In the same way, the feeling that an extreme show of affection is undesirable becomes progressively entrenched, and while all this is inculcated in small doses, the process begins in those early stages. He will put up with any pain and cultivate a will of iron to avoid showing his feelings. These are the conditions for being a good soldier and a good commanding officer.

We were in the office of the Commander of the Army Academy.

An adjutant came in and said quietly: 'They are ready, Sir!' I could not quite understand what was going on. The commander said: 'Right, bring them in.' I was a bit taken aback when the chief cook entered, wheeling in a trolley bearing dishes of food. It was, after all, only eleven o'clock in the morning. The commander saw my surprise and said:

'Don't worry, it's not for you. It has been brought for me to taste.'

I was no wiser.

'What will you taste?'

'Every item of food served at the school has to be tasted by the commander or, in his absence, by his deputy. The samples are then refrigerated and kept for twenty-four hours.'

It had never occurred to me that a commander could be used as a guinea pig. It turned out that this is one of their proudest traditions. The commander is expected to be satisfied that the students are served proper and wholesome food. I could not help making a mental comparison with the civilian schools I have known.

On arrival at the college the students are on the whole undernourished, an accurate reflection, no doubt, of Turkey's general nutrition problem.

A look at the background of the majority of the children is enough to show that their priority has been filling their stomachs rather than obtaining nutrition. This is supported by their real state of malnutrition when they arrive here. To start with, we provide a special diet to overcome this. After that, we provide a regular diet of 3,500 to 4,000 calories a day. The comparable diet outside the school varies between 1,00 and 2,500 calories. We are trying to cut down on carbohydrates as much as possible, but established habits die hard; every time we cut down on bread, they complain that they don't feel satisfied.

In the military schools every meal is properly thought out, complete with its meat, vegetable and sweet courses. But the most important fact is that, with proper meals and mealtimes, the children make up for lost ground in a few months.

I had never eaten such food in my whole life until I came to this school. Meat was served only once a month at home. We mostly fed on soup thickened with bread.

Another 'first' is personal cleanliness.

We had hot water once a month at home to wash ourselves.
Here at the school we not only have our weekly session at the
communal bath but we take a shower without fail after each
running or PT session. Some of us were even a bit put out by
this to start with but, of course, one quickly gets into the habit.

In the dining hall, the students lined up, ten to a table. All eyes
were on the School Commandant. They stood to attention.
Prayers were said in unison. 'We thank God. May our nation
live for ever.' The Commandant added: 'Enjoy your meal.' The
answer came in unison: 'Thank you, Sir.'
At each table, one of the students served the food. It struck
me that almost every single student ate the same way, with knife
and fork. I was curious how this had been achieved.

The first thing we teach the boys when they come here is how to
eat their food, how to use a knife and fork, how to drink water
without slurping and belching. Naturally, some have no need to
be taught table manners but we go through it with all of them.
There are some who have never used a knife and fork before
or eaten off a plate.

There is not a single trace of loud talking or pushing and
shoving at the table. The food is eaten as though they were all
very serious adults; only their bright eyes rolling mischievously
reveal that these are youths of only fourteen to eighteen.
At the end of the meal grace is repeated. The old custom
whereby we did not get down from the table before our
seniors is in force at this school. Everybody stands to attention,
until the Commander leaves the table and the hall, nor does
pandemonium break out when he is gone. They disperse in
groups of two or three, carefully keeping their voices down.

*   *   *

If boys at the schools find this discipline uncomfortable, once
they pass on to the Military Academies they quickly realize
what a picnic these first few years have been. At school level

care is taken not to pile on the pressure too much. The aim is to see the very young treated with greater tolerance than is shown at the Military Academies. Those at the Military Schools have another advantage. If they wish, they can pay a certain compensation to the state and leave the school. This is an advantage they will not easily enjoy once they enter the Military Academy. Nevertheless, they receive training throughout their school years to develop their social life, for example, how to go to the theatre for the first time, how to applaud, and how to enter a restaurant and order a meal.

Admission to the Military Schools and Academies is extremely difficult, so meticulous is the selection process and so great the efforts 'to select the best'. But, in the final analysis, is it the *crème de la crème* that is selected? This is how the Commander of a Military Academy sums up his experience:

> No. First, we have to be satisfied with the standards we set. We can't expect more. Our greatest problem is the raw material we receive. Despite our efforts to choose the best available, we manage to do so in some areas but not in others. Think of the difference between the best from Istanbul and the best from Nevsehir and you'll know what I mean.

To gain admission to the Military School an applicant must prove that he did well in science in junior school. He will apply to the Military School with the marks he scored in a test taken at a science college. He has no chance of admission if his marks are below a certain level. The applicant with sufficient marks first submits to an oral interview.

> This first stage in selection is mainly to find out if the applicant knows Turkish and whether he speaks it well or badly, but it is also to check his appearance. Well, I mean, we have had stutterers, some with huge boils, some with asymmetrical faces and others with flat feet. Sometimes they're bald, or colour-blind, or cross-eyed, or dwarfs. We get rid of them straightaway. We are very keen to have people with as normal an appearance as possible. It is not that we are seeking good looks; we just do not want anything abnormal.

Attention is paid to weeding out those who have had serious illnesses, including heart problems, and those who are excessively fat or thin. The candidates who meet these criteria undergo a physical examination and physical tests like running 1,600 metres in a given time and doing a certain number of push-ups, and so forth. Usually by this stage 20 per cent of the applicants have been eliminated. Finally there is a general knowledge test. The points gained in this test, which is set at the level of the state colleges, are added together. If a total of, say, 1,200 students are to be admitted in a given year, then the 1,200 applicants with the highest scores are admitted.

That, however, is not the end. The vetting then begins. Does the applicant have a police record? What are the occupations of his father and mother and do they have a good record? Investigations are carried out to find out if any members of his family have been involved in political incidents, or have been convicted of left- or right-wing activities, or of theft or fraud. Some of these investigations are based on records, the rest on inquiries by the local police of the butcher or the greengrocer.

> What else can we do? These are the means at our disposal. In some cases we find that the parent has a criminal record and we immediately reject the candidate. An ideological streak even in a very distant relative affects the child.

The greatest problems, though they are few, concern the students' connections with their families, near or far. The youth's own ideas are taken seriously but it is sometimes concluded that he is still influenced by his family and this results in his dismissal.

> Towards the end of my first year, I was summoned to see the Commander of the School. I knew immediately there was something wrong. When I was ushered in, the Commander had a civilian opposite him. The Commander was annoyed. He said, 'I understand you have a cousin called Ahmet'. To tell the truth, I did not quite know who he was talking about. I did have a cousin I had never met in my whole life, and whose name I had heard once. I did not even know what he did. He lived in Adiyaman. When I told the Commander I didn't know him, he was even more annoyed and shouted, 'Why did you not tell me that this

man was a trade unionist who went to prison for inciting strikes?'
I was quite unable to explain that I did not know my cousin at
all, and I was eventually expelled from the Military School.

The commanders admit that the selection process is very
carefully considered, but they say no action is taken without
good additional reasons.

The incident you refer to was not all that simple. We take no
action until a close investigation makes it clear that the individual
in question may not be fit to enter the Military Academy; we
have had bitter experiences in the past and have to guard against
extremists.

Now, could this be the last stage in the obstacle race? Not yet!
The applicant's next step is to go to a fully-equipped hospital and
obtain a report that he is 'fit to be a soldier'.

If he survives all this, then and only then can he officially
register.

While making a tour of the Military School, visiting the
classrooms and talking to the students, one quickly senses that
these are 'precious' students. One is struck by the number of
facilities and the attention lavished on them, which cannot be
matched in most civilian schools.

The first thing that struck me was the size of the classes. In
civilian schools, each class averages sixty to seventy students.
Students practically sit in each other's laps and it is hard to
tell whether a teacher is teaching or a public debate is in
progress. I took a count while touring Kuleli Military High
School and found that no class had more than twenty-five
students. Everything is spick and span. There is a gigantic
library, panelled to the ceiling, with 25,000 books; language
laboratories galore; chemistry labs with plenty of new equipment
– these are things that one could not find even in the most
select fee-paying schools. With its electronics and computer
department, Kuleli has the facilities to train the students to
full Western standards. Students in Turkey's civilian schools
are unable to enjoy extra-mural activities because of the lack
of resources; but those at Kuleli produce their own films and
TV programmes, their own monthly publication printed on

their own press, and train in thirteen Olympic activities, in sports facilities that would be the envy of professional sportsmen.

What impressed me most was the degree of close attention paid to the individual students. An original system, unparalleled in any civilian school, has been developed and is in force in all Military Schools and Academies, and in the NCO schools. Shortcomings in training are constantly followed up and exposed by an 'Assessment Department'. Examination papers are processed by optical readers and the data fed into computers. This shows up lessons that are not going smoothly, or reasons for students lagging behind, or material they do not understand. The results are fed into the 'Question Bank', which contains several hundred thousand questions, and enables the teachers to carry out a test at any level they choose. It also provides an indication of exactly how difficult each question is.

Another practice is the testing of teachers. Questions are obtained at random from the question bank and put to all teachers of the same subject at the same level in various military schools. Thus it is possible to keep an eye on the success of teachers and also to standardize their teaching. Civilian schools have no such system; nor is it contemplated.

We study every student from all aspects. We carry out two surveys a year and follow them through to the final year. We question them on all subjects and feed the results into the computer. When the commander asks a question, all we have to do is press the key for the answer. This information follows the young officer all the way to his promotion to full general.

What do the surveys reveal? The ability to follow orders; the ability to understand social concepts; imagination and creativity; the ability to see relationships in complex situations; the ability to observe and analyse; reading aloud with understanding; reading silently with understanding; manifest adolescent problems; phobias; irritability, shyness, egotism, aggressiveness, jealousy, bad habits, breaches of discipline, possession of ideals or lack of them, incidents of lying and thieving, working habits; the students' major interests; his leadership qualities. Is he easily influenced by others? Does he show his emotions? etc. etc. This continuous assessment from the age of fourteen to eighteen shows who is going to be a good officer and what his prospects of promotion are. Their margin of error is under 5 per cent.

A questionnaire of 200 queries is used to probe the darkest and most secret recesses of a student's mind. Are the classrooms noisy? Can the student work properly? Are the meals satisfactory? Which lessons does he find hard going? Does he think some students get special attention? Are the lessons sufficiently interesting? Does he get nervous during examinations? He is then asked about his family, his social adaptation, his interest in the opposite sex, his emotional life, and everything to do with the future.

When the time comes for a student to leave school, and the computers collate all the surveys, they reveal a complete picture of every student's prospects.

In one of the questionnaires, a student was asked where he wanted to spend the weekend, at school or at home. He replied, 'at school'. During a survey of the answers, this one immediately attracted the attention of the tutors. You might think their reaction would be, 'Bravo, look how he loves the soldier's home from home', or something similar. But, on the contrary, this boy was singled out for attention. After all, it was abnormal for a sixteen-year old whose home was in Istanbul to want to spend his weekend leave at school. The matter was looked into and it was discovered that this boy's parents were in the throes of a divorce and were constantly at each other's throats. When we discovered this, the boy was provided with a counsellor to act as a supportive elder brother until matters were sorted out. This is not a one-off case. Students are individually monitored like this, and treated according to their circumstances.

The continuous assessment, orientation, filling-in-the-blanks on who's who and on professional choice, all go to complete the picture.

When I went to Kuleli to see what went on there, I had the idea that children were being trained in weapons from a tender age. The truth is that the Military Schools follow the four-year programme of the state colleges, where courses are taught in a foreign language. That is to say, the programme of the Ministry of National Education is applied without modification. There is only one difference: the programme is taken much more seriously. Along with the normal college education, the students are given a 'superior physical education' and what amounts to 'a basic military training'.

All these are followed through within a given disciplinary and systematic framework. Thus, for instance, the lessons on 'Ataturkism' are the same in civilian and Military Schools, but in the latter they are taught with more emphasis. For the past three years the teaching of foreign languages has been a priority subject in military schools.

> The scholastic performance of every student is constantly monitored. If he is lagging behind in English or any other subject, his weekend leave is immediately cancelled, he is made to do extra studies and is supervised. We keep track of every student's studies better than they do themselves. We make them work and instil a sense of discipline.

The students undergo tests to determine their stamina and aerobic capacity, to establish their aptitude for various branches of sport, and to decide which branch they should specialize in. The month-long summer camp is devoted to sports and military instruction. As 60 per cent of the intake cannot swim, swimming classes are the most important. Except for the weekends and a forty-five-day summer holiday, the students are kept constantly busy.

This kind of intensive attention, aided by surveys and computer appraisals, of course results in high average success rate – actually 97 per cent.

> If we conclude at the end of the first year that some students will not make the grade despite all our devoted efforts, we throw them out. As a result we end up with a higher success rate.

When I investigated whether another reason for this success lay in high expenditure, I discovered that the Military Schools achieved greater productivity not by drawing on substantial material resources (though it should be said that their resources are not scant) but through the systematic use of what is available, by a superior definition of their priorities, better organization, and the knowledge of their needs one year in advance.

What else is available to the student?

Going through the corridors, I came across a twelve-point list on a pocket-sized card in the hands of one of the students. It was to be kept with him always, to be read and learnt by heart. It described, point by point, what sort of person he should aim to be. It said:

(i)   I shall be industrious.
(ii)   I shall be honest, always telling the truth even though it may hurt me.
(iii)   I shall cherish the profession of soldiering and encourage others to do the same.
(iv)   I shall uphold the honour of my uniform by my behaviour.
(v)   I shall avoid arrogance but be dignified.
(vi)   I shall keep myself and my environment clean, and preserve it.
(vii)   I shall compete not with others but with myself. I shall adapt to my generation through study and work, and I shall always strive to improve.
(viii) To the respect I feel towards my seniors and commanders I shall add affection towards my juniors.
(ix)   I shall use every opportunity to develop myself physically by engaging in sports.
(x)   I shall protect myself from every harmful habit.
(xi)   I shall develop moral qualities.
(xii)   I shall find out my weaknesses and overcome them.

Reading this card reveals that the Military Schools strive to eliminate from the young people who have come into their hands all the faults of Turkish society, and to recreate them as ideal Turkish youth. Are they successful in their efforts? I recall the words of the Commander of a Military Academy:

> Irrespective of the excellence of the education you provide, with the clay in your hands you can only go so far and no further.

All efforts are directed, from the school level onwards, to teaching the future commander all he needs to learn, from

Ataturkist principles to the defence and protection of the Republic.

We got together with a group of thirty students from three military schools. They had all reached the final year. It would not be long before they completed their studies and went their separate ways to one of the Military Academies. So here I was, facing the future core of the Turkish Armed Forces, all so young, so spotlessly clean.

'Now then, lads. Let us begin, if you like, by getting you to ask me anything you wish. Then it will be my turn to question you.'

So, for an entire hour, they grilled me. Some of the questions thrown at me were of a very high order indeed. The student of yesteryear was gone, transformed into someone entirely different.

When my turn came, I began with the questions that interested me most.

'Lads, do you believe the armed forces have a right to intervene if the need arises?'

Their faces displayed amused incredulity. I had expected them to avoid answering. After all, the Commander might get to hear what was said and might be annoyed, or a situation might arise which would be open to misinterpretation. On the contrary, they seemed to regard my question as a bit odd, as if they thought, 'How on earth can he not know the answer?'

'Sir, we are the army of the regime. It is our duty to defend and guard this country, and to keep the state sound and the regime secure.'

'Do you believe in democracy?'

'Of course, is there a better form of government?'

'Fine. But what happens if a government elected by the people's votes takes a decision that you consider to be against Ataturk's principles; do you still consider that you have the right to intervene?'

A hand shot up from the back of the hall. I received a reply in a firm voice. 'We are opposed to anybody, no matter whether they are there by the grace of the ballot box or the votes of the National Assembly, who attempts to violate Ataturk's principles. We have a right to act to this end in the interests of our people, and for their protection.'

'All right, lads, do you believe politicians?'

'No. They may lie and deceive the people to suit themselves. I have yet to hear anything favourable about politicians.'

'Where are you taught that politicians are not to be believed – at school?'

'Not from the commanders but more from my father. Newspaper articles, too, say that politicians always work in their own self-interest.'

'You look and sound as if you do not wish the politicians to govern this country. Well, wouldn't you like to govern yourselves?'

They laugh in unison and utter a drawn-out 'No! That is not our job.' Then they add immediately, 'If any politician attempts to embrace attitudes that run counter to our concept of Turkey, the Turkey we believe in, he will find himself up against us.'

'But, lads, don't you have faith in the people? You are, after all, talking about those chosen by the people.'

'My people are poorly-educated. They may be misled by politicians and the self-interested.'

'So you see yourselves as Turkey's guardians.'

'Of course we do. As long as we are surrounded by external enemies, and face perverted ideas at home, we are the country's guardians. We need, therefore, to be very well-prepared.'

'Who has given you this task?'

'Ataturk and our elders. We have been told this is our right, even our duty. And we have accepted it as such.'

'When do you think you may give up these ideas?'

'The day the civilians and politicians mend their ways.'

The average age of my interlocutors was eighteen. A year later, they would be on their way to the Military Academies.

3
Shirts of Steel: The Military Academy

> I am not asking you to train doctors or engineers; I am asking you to train team leaders, officers with leadership qualities . . .
>
> Chief of the General Staff

One soon finds out that life in the Military School was something of a picnic, compared with life in the Military Academies, the Turkish equivalents of Sandhurst, Cranwell and Dartmouth Naval College in Britain.

A student's years at the Military Academy begin with the oath-taking ceremony, which is most impressive and colourful and the first step in the military career of a future commander. Those taking the oath gather round a table covered with the Turkish flag, with such weapons as a mortar and a machine-gun on it, placing one hand on the flag and the other on a weapon. This is their first step into a world from which there is no easy return.

> I swear on my honour that I shall serve my nation and republic at all times with loyalty and devotion, in peace and in war, on land, at sea and in the air, that I shall obey the law, regulations and my superiors, and that I shall hold the honour of the military profession and the glory of the Turkish flag dearer than my own life which I shall willingly sacrifice, if necessary, for the country, the republic, and my duty . . .

The commander makes a speech, partly to explain the meaning of the acts in the oath-taking ceremony:

In putting one hand on the shoulder of the comrade-in-arms next to you and holding tight to him, you will demonstrate your unity and solidarity and show that you love your profession more than your own life, and that you are bound to your comrades-in-arms and to the Armed Forces by an unshakeable faith. You will place your other hand on the weapons laid out on the glorious Turkish flag, and you will loudly proclaim that you will put the interests of the country and the nation above all else and, when necessary, you will not hesitate to shed your blood and lay down your life in their defence, and you will give proof of the traditional courage of the Turks. Your upright bodies will represent the strength and will of a soldier brought up in the spirit of discipline, a soldier who is obedient, possesses a powerful sense of morality, is honest and reliable and is imbued with the love of the country and the nation.

A large proportion of the Military Academy students I talked to after the ceremony did not know precisely what kind of life they had embarked upon or what lay ahead. They were just happy to have come through an extremely tough selection process and to have reached a stage keenly awaited by their families: 'I'm very proud I'm going to be a soldier'; 'I have achieved my ideal.' If one examines these answers more closely, the fact emerges that a substantial majority are happy 'to have found a reliable and lasting job, to have secured a future, to know that they will no more be a burden to their families and that they have achieved a better quality of life than before'. There was not much evidence of the kind of reasons you would expect from entrants to the air force or naval schools, like 'fascination with flying, a desire to master technology through becoming an airman, or the irresistible attraction of the sea'. This is a normal response if one considers that 70–80 per cent of those admitted to the Academies come from small-town families with three to four children and monthly incomes of 30,000–70,000 TL.

The intake of the Military Academies is partly from military high schools, partly from civilian schools (some 40–60 per cent). Those from the Military Schools make a more informed choice and are more ambitious. The entrants from the civilian side, like the new students at military high schools, are in for a shaky start – the intense loneliness of the first few months, confusion arising from educational deficiencies, and the experience of wholesome and balanced nutrition.

Once the ceremonies are over and the school doors shut behind the students, a gigantic piece of machinery goes into action. It takes the undeveloped youth who is not quite sure of what he wants – who has chosen the Military Academy under the guidance of his family, who has come with the customs of a poor or middle-class background, its social traditions and national and world outlook – and transforms him into a totally different person.

For the same age group, another piece of machinery goes into action in civilian schools and universities but, at the end of the four-year period of their training, the two sets of individuals turned out by the civilian and military machines are totally different from each other. How does this difference arise?

As you tour any of the Military Academies, you can immediately see that the source of the difference lies in discipline. Discipline pervades everything; in no way can a single line of the instructions issued be transgressed. Watching the students going quietly around the parade-ground in groups of two or three, witnessing the orderly manners at mealtimes, the tidiness of the dormitories, their behaviour in entering or leaving the classrooms – in short, sharing their daily lives – I came to have a better idea of what the 'steel shirt' was all about.

A substantial majority of those entering the Military Academies encounter 'first time ever' experiences and have to make some major adjustments.

While staying at one of the Academies, I was awakened by some noises in the exercise yard outside. A cold rain was beating down in the darkness of the early morning. The time was 6.30 a.m. Wondering if anything untoward had happened, I went to the window and could not believe my eyes. Wearing tracksuits, the students were going out on a run. They had got up at 6.00 or 6.15, had a wash, and were ready to go. The regimental commander smiled at my obvious astonishment:

> They have a 5-kilometre run at this hour every morning. That lasts until 7.30. Then they have a shower and go down for breakfast at 8.00. They have to cover the run in a given time . . . After all this physical effort, four consecutive hours of classes begin at 8.30 and

end at 12.30 when lunch is served. They have only one hour for lunch. They go into the dining-room together, take their assigned seats and wait for the arrival of the most senior officer. There are prayers and then eating the meal in the prescribed manner.

At each table, students take turns at serving. No restriction is placed on conversation but voices have to be kept low. The students leave the dining-room after the commander. There is no such thing as a fast exit by anyone who has finished eating. Leaving is also properly regulated – in groups of two or three, at a steady pace. I made a mental comparison of this with my own student days at Galatasaray Lycée (in Istanbul): the stampede into the dining-room to gobble down the food, then the rush back into the yard to kick a ball around madly, the return to classes in the afternoon totally exhausted and dripping with perspiration. The difference was instantly obvious.

That is not all, however.

When the classes from 1.20 to 3.30 p.m. are over, the students have a mere half-hour to get themselves ready. Group military physical exercises begin at 4.00: pentathlon, full-pack drill, bayonet exercise, and so forth. As these PT exercises end, the training of those selected for school teams begins between 5.00 and 6.30. The rest engage in educational activities such as music, drawing or using computers.

The only uncommitted time of leisure for the Military Academy students is from 6.30 till 7.00. This is when they can have a few moments to rest, to collect themselves, or have a few words with their fellow students. Dinner from 7.00 to 8.00 is taken in much the same way as lunch and then comes a study period until 10 p.m. The students have two hours to review their lessons of the day and to prepare those of the next – provided they manage to fight off fatigue and do not doze off. Lights-out is at 10.15 when all noise ceases.

The steel shirt comes off only in sleep.

The student of the Military Academy spends four entire years in this way. We shall come back later to what he is taught, but even this much leaves one breathless.

At weekends, except for those whose homes are in the same city or very close to it, or who have relations to visit, students have to stay in the school. In the summer, there is training in sports. Students are entitled to only forty-five days off a year to spend with their parents and brothers and sisters whom they have been dying to see.

What sort of officer do the Chiefs of the General Staff want these schools to produce? To be more precise, what kind of commander are they aiming at?

To begin with, special attention is paid to ensuring certain fundamental differences between the Ottoman officer and the officer of the post-1923 period. The intention is to inculcate in these young students the spirit, traditions and moral values of the officer in the period of Turkey's War of Independence.

The education that the Military Academies aim to give the young commander focuses on five points:

(i) Military skills: To ensure that he has the knowledge, physical toughness and stamina to train and lead a unit under his future command.

(ii) General culture: To give him a good command of a foreign language, to bring his cultural standards up to the level required by the modern world, particularly when it comes to understanding world affairs, and to enable him to set his thoughts down in writing.

(iii) Social side: The making of a courteous and gentlemanly individual who will know how to behave in any circumstances, will be an orderly, decent, honourable man knowing how to gain acceptance in any assembly, protecting the honour of the uniform he wears, and devoted, in the fullest sense, to the Armed Forces.

(iv) Discipline: He will not disobey his commander or his orders; he will obey without question and will also be extremely hard-working and self-sacrificing.

(v) Ideology: Just as he will protect the country against external threat and be well versed in the art of war, he will also be on the look-out for dangers from within the country. He will be full of love for the country and the nation, and will reject any ideology outside Ataturk's principles, instantly identifying and opposing separatist and extremist movements.

The Academies are trying to create an entirely different kind of Turk from the raw material they get, an ideal Turk, the kind of Turk one dreams of. He is free of all the maladies thought to afflict Turkish society; he is extremely well-informed, trustworthy and has all the social graces; he is also a proud and honourable warrior, a man of discipline and integrity.

Is this a difficult goal to achieve? Let the Commander of one of the Academies answer this question:

> We know the extent of the resources of Turkish society and that the raw material is not the raw material of America or Europe, but our own. We try to get the best out of what is available to us and in some cases we achieve our goal, in others a parody of it.

To achieve the General Staff's long list of expectations, the schools follow an intensive course. There is the physical training and the input of knowledge and, moreover, something that other armed forces do not practise but which is of special importance in the Turkish forces: lessons in ideas and morale. In addition, the Military Academies strive to wean their young charges from the conditioning of their own background and to train them to participate easily in the social life of Turkey's major cities and the capitals of Europe by constantly teaching them etiquette and social usage.

Another area which takes up a lot of time and effort is how to endow the young student with 'initiative'.

> Some are born with initiative. They are gregarious and have strong characters. Others are inhibited. The youth of our country are brought up by their families in the spirit of 'don't-do-this, don't-touch-that' and a beating is the fairly common method of obtaining compliance. These are the youths we are trying to help achieve 'personality'. These are the youths we are teaching to have initiative. Naturally we succeed in some cases, but fail in others.

Hearing the Academy Commander, I kept making mental comparisons with educational practices in the civilian high schools and universities, with similar problems and corresponding approaches to the same questions.

Admission to the Military Academies is not so easy as to the military high schools. Anybody under twenty years of age who has completed a lycée with a science or modern curriculum and foreign language courses may apply for admission to the Military Academies, provided he has scored at least 140 points in the first stage of the university entrance examinations. So, from the very beginning, candidates above a certain standard are given preference. Those below it would not be able to keep up with the pace of education at the Military Academies. The applicants have to undergo an interview (a personality-evaluation test) followed by a physical examination. As in the military high schools, any possible abnormality of personality in the applicant and any excessive shortcoming in his physique are checked and the first selections are made. The very short, the excessively fat, the hunchbacks, the flat-footed, the colour-blind and those with attributes incompatible with soldiering, as well as those with a legacy of a serious illness in their past, are weeded out. The candidate for the Military Academy will also have to pass a number of sports tests, as he did to gain admittance to the high school. They are then required to obtain a fitness report from a fully-equipped hospital.

The second stage of the university entrance examinations is held in May. The exam for the Military Academy is in June. By the time the candidate comes to register in the first week of August, the Academy administration has been supplied by the University Placement Centre with the results of the second stage of his university entrance exams. If he has gained a place at a university, he is informed by the Academy. If he finds that he has won a place at a good university and can study his own choice of subject, he usually opts for the civilian life. Nevertheless, quite a few choose the military.

Since at least one person from every province in Turkey *must* be enrolled, it becomes difficult to claim that the Military Academies are able to attract the *crème de la crème*. Despite the selection of individuals above a certain standard, it is generally young men of average ability who join up.

There are two more obstacles still awaiting our candidate officer. The first is the confidential investigation to be carried out about himself and his family. The second is the one-month adaptation course that the candidate will undergo before the academic year begins. This is a period when school conditions are applied in full and ends with some weeding out among the candidates. Once the oath-taking ceremony is over, the future commander is in school. He may not receive visitors or go out during week days. The steel shirt is ready to be worn, and there is no easy way to take it off.

> We have to take this young man entrusted to us and completely REMAKE him, in four years. The way to achieve this is by adjusting the existing educational system to our needs and applying it, almost forcibly, through extremely tight discipline.

The present educational system in the Turkish Military Academies embodies the changes wrought by Truman aid to Turkey, Turkey's involvement in the Korean conflict and its entry into NATO. Until then, there was an entirely different system in force. This is what a General who graduated in 1940 says about his own time as a military student:

> If someone failed twice in three years in the Military School or Academy, he would be immediately reduced to the ranks. On top of that, half of the period spent in the Military School or Academy was added to the time he had to serve in the ranks as a conscript. Fear of all that made us learn everything by heart.
> We were completely cut off from the world. We were brought up entirely within our own circle. Forget about libraries and the like; there was even a ban on reading newspapers. We just had an intense military training. The German educational system had been taken over as a whole and put into practice with certain changes. There were no ideologies to speak of in Turkey's political and social life in those days. There was not much talk about Ataturk's principles. It was, anyway, the single-party period and Pasha Ismet Inonu was the leader. There was no need for the Ataturk banner. One knew what the 'six arrows' of the Republican People's Party stood for, and Ataturk's Great Speech was required reading, and that was all. The most important thing was loyalty to the state and the government. Scrupulous attention was devoted to training the kind of officer who would be obedient to his commanders and to the leaders of the state.

While the system has been Americanized from the 1950s onwards, it is founded on German, or even old Prussian, principles: absolute loyalty to the motherland, rigid discipline, blind belief in the commanders and unquestioning obedience. Despite the fact that Germany relinquished the Prussian system after the Second World War and became Americanized like Turkish civilian society, the Prussian approach in the Turkish Army has not yet been erased.

The system currently in practice was derived by adapting American methods to Turkish conditions (more correctly, by translating them into Turkish terms) and is peculiar to Turkey. Moreover, a considerable amount has been changed by adjusting – once again, according to the needs of the day – the ratio of the subjects in the curriculum.

The General Staff is the sole judge of the subjects to be taught and how much time should be devoted to them, and what textbooks are to be studied in the Military Academies. The Ministry of Education does not see these books and has no idea of their contents or in what way they might differ from their civilian counterparts. If the Ministry had been interested, copies would have been sent, but no Minister of Education has ever felt it necessary or been bold enough to ask, or to intervene if need be.

In the curricula of the Military Schools in the 1950s the emphasis was on three military subjects; 70 per cent of the classes were on military subjects and 30 per cent on academic ones. In the atmosphere ushered in by the military coup of 27 May 1960 which toppled the Menderes government, it became easier for liberal ideas to enter the schools and the courses at the Military Academies were extended from three years to four. After 1974, the programme of the Middle East Technical University (in Ankara) was put into practice; this time, 70 per cent of the classes were devoted to academic subjects and 30 per cent to military ones.

This change, however, soon produced a major difficulty for the General Staff. Newly-commissioned second lieutenants were hesitant in their first command because of the inadequacy of their military training. Moreover, with the entry of ideologies into the Military Academies problems began to arise. The situation in the Academies operating under this programme around 1978, when political violence was at its height, alarmed the General Staff. Soon afterwards, the proportion of classes was changed to 56 per cent military (including sports) and 44 per cent academic. A definitive barrier was raised against liberal tendencies. Although teaching at the Military Academies is to graduate level, legally students are not considered university graduates. The Higher Education Council regard their diplomas as equal to those issued by the institutions of higher education, but because of the rapid and efficient pace of teaching at the military establishments, many of their students easily reach the standards achieved at a number of universities. The Air Force Academy teaches about 60 per cent of what is taught to a civilian student at the Istanbul Technical University. Because of the different requirements of the service, the education at the Naval Academy is different from that at the Military Academy.

We were struck by the inclusion of political and economic subjects beyond the requirements of a modern officer, and the efforts made to train 'managers'. One gets the impression that in recent years there has been an increase in the teaching of socio-political and economic subjects which are necessary for 'managers' rather than for officers directing troops.

The military topics (like tactics, logistics, topography, armoured units, intelligence, artillery, supplies, military geography) which account for 56 per cent of the classes are limited to the use of equipment and supplies currently possessed by the Turkish Armed Forces. As yet, there is no training in the high technology of the industrialized countries. It follows that military instruction is limited by the facilities available. New developments and the effects of high technology on weapons systems and on command and control are closely followed, but little can be done about them under present conditions.

To ensure the proper training of the future officer, the state is straining its resources almost beyond the limit. While the same cannot yet be said for the Army Academy (Harbiye), the naval and air force equivalents are much above Turkish standards. The Naval Military Academy in particular has sections superior to a number of corresponding schools I have visited in Western Europe, sections that may become inadequate in the future, but which strike today's observer as de luxe. Perhaps we may not make this judgement in fifteen years' time but, having seen the conditions for students in Turkey's civilian schools and universities, one cannot help feeling uneasy at such a great difference in standards between the two.

The outstanding feature of the Military Academies is their superior educational facilities. For example, while classes average 300 at Turkish universities, the comparative figure in the Military Academies is 10–20: the aim is to have one teacher to every ten students. A substantial number of classes are held in English, and a second foreign language is considered essential. There is a choice of ten different languages, including Russian, and teaching methods follow the most up-to-date audio-visual methods.

Vocational classes, too, have far more laboratories and equipment for experiments than the civilian schools and universities. To help raise the general cultural level of the student, the Air Force Academy has one library for every sixty-five students. Nor are the teachers overwhelmed by their commitments – their load is no more than 380 hours teaching per year, or twelve hours per week. The dormitories average four persons to a room. The Naval Academy has a gigantic cinema hall holding 2,000 and conference halls of various sizes, and the Air Force Academy boasts indoor sports arenas and swimming pools unequalled in Turkey.

The food served provides 4,000 calories per day and is of extremely good quality. Finally, each academy has an increasing number of computer units, closed-circuit TV systems vying with the Turkish Radio and Television Corporation in the modernity of their equipment, and accompanying common rooms and billiard rooms.

There is no way one can make a complete inventory of all these facilities. It can only be said that the state has provided, without exception, every modern comfort and educational facility necessary for the budding commander, and that there is nothing comparable in any civilian educational establishment. The cost of training these budding commanders is, as one might imagine, considerable. For instance, a second lieutenant who graduates from the Army Academy costs 7 million TL. In the case of the navy, the figure rises to 23 million. As everywhere else in the world, the most expensive person to train is a pilot officer, at 184 million TL.

As in the Military Schools, the students at the Military Academy are the object of dedicated care and attention. Every minute of their time is under scrutiny and they are not allowed to have an idle moment. Their private problems, family life, relations with fellow students, and personal interests are constantly under review through a regular series of tests. If these reveal any hint of trouble, tutors immediately help to resolve their problems. If a student is found to be lagging behind in any subject, attention is focused on him to pressurize him to close the gap, and weekend passes are cancelled. Each student's personal record card depicts a precise chart of the ups and downs he has experienced from the moment he set foot in the academy; the result is a 90 per cent success rate.

Within this framework, the budding commanders are encouraged to compete.

> If you lag behind in your lessons or get a bad mark in your record, the man next to you may get promotion on the day you get your first assignment and end up as your commander. Just bear that in mind.

This is the most effective spur for each prospective officer who thus finds himself caught up in a hectic race.

> You must be the best; you must get the best grades for knowledge, experience and progress.

Over four years this daily exhortation sets the nerves a little more on edge with every passing day. Moreover, the students at the Military Academy are in bondage: there is no dropping out halfway through. It is just not possible to reach the second or third year and then throw in the towel. A student may not buy himself out as one can in the Military Schools. The Military Academy law of 1962 will not allow it. If he commits a crime, gets married, or fails twice in succession to be promoted, he is expelled. The Ministry of the Interior, the universities and other organizations are immediately informed of his expulsion to prevent his admission to any state organization. In other words, the student who enrols at the Military Academy has no alternative but to participate in the race and see it through. This too is a factor that naturally increases tension.

There are five offences the Military Academies will not tolerate: violation of discipline, failure to get promotion, involvement in ideological or party activities, larceny, and homosexuality. There are heavy penalties not only for the offenders but also for those who witness or are aware of any of these offences and fail to report them. Such failures may result in expulsion.

During the 1977–84 period, the following were expelled from the Military Academy: 318 for indiscipline, 100 for failing to be promoted, and 22 for health reasons. (This was in addition to the 224 who left during the one-month trial period due to incompatibility or at their own request). In the same period 112 were removed from the Naval Academy: 51 for indiscipline, 51 for poor performance and 9 for incompatibility. The total for the air force was 71.

There was also a substantial number of further expulsions for ideological reasons and involvement in 'incidents'. In view of the fact that the Military Academies do not admit students from the Theological Schools, the term 'ideological reasons' as used here means, in the main, having left-wing sympathies or supporting those with interventionist inclinations. The role played by the Army Academy (Harbiye) in the 27 May 1960 coup that toppled the Menderes Government, and the two abortive coups in 1962 and 1963 led by Colonel Aydemir, are still regarded in the higher ranks of the army as disturbing developments, and it is well-known that in all periods of unrest there are expulsions from the Academies. The biggest such occurrence on record is the expulsion of 1,400 Army Academy students for their participation on 21 May 1963 in the second attempted Aydemir coup. As a result of these purges there were no graduating classes from Harbiye in 1963 and 1964. Between 1977 and 1984, too, some 1,200 students and graduates were weeded out from the Army Academy.

It would be no exaggeration to say that the strictest watch kept is on ideological matters.

In touring the Military Academies, one cannot help wondering if these young students are not being overloaded. The truth is that teaching is done under remarkably high pressure. But how much of it is actually absorbed and how much is learnt by heart just for the exams and then forgotten? The answer I always received was:

> We have to impose this overload because we are training officers for the army of a poor country. Our officer cannot specialize in a narrow field and be confined to carrying out a given limited task as in wealthy Western countries. He has to be something of a general practitioner. He must be able to tackle nearly everything at one time or another, from teaching the raw recruit how to open his legs and relieve himself, aiming for the dead centre of the pit below him, to knowing how to operate the most modern weaponry. Our officers will deal with computers one day, supplies the next, and strategy the day after. We are not rich enough, unfortunately, to train officers to know only one branch extremely well.

But is this not providing them with superficial knowledge?

> How much does he absorb? Of course, not all. Some still have that vacant look in the eye in their final year. Some get there in the end by dint of rote. Some of my charges absorb the knowledge given them only superficially. But the tests show that 40 per cent of the graduates of the Military Academies will go straight to the top. The remaining 60 per cent get by. Now, if we could get this kind of result in all schools and universities Turkey would progress.

A short study I conducted at the Military Academies immediately disclosed that the basic problem arose from the education in the schools. The intake from the Military Schools adjust to the training at the Military Academies without any difficulty, but there is a great deal of confusion among those who come from the civilian schools – the reason lies in the low standard of education in such establishments. In the Naval Academy's first year of 238 students, it was found that forty-three were weak and three-quarters of them had come from civilian schools.

Another subject that prospective commanders are taught from the earliest stages is 'situation appraisal'. This can also be called 'methodology of logic', or 'a method of clear thinking and drawing organized conclusions' that they will apply till the end of their lives, in military, social or political matters, and particularly in their personal lives.

They are taught a prescription for clear thinking – how to make an appraisal when faced with a situation, how to make a decision, and how best to act on that decision. You hear from time to time or learn from the papers that officers have come together to 'evaluate a situation'. Sometimes, listening to their conversation, you are aware that almost all their thought processes touch on the same area. This does not mean that all officers think the same way, but it does show that 'they think within the framework of the same plan'.

Command situation appraisal consists of

(i)   The situation (the prevailing conditions).
(ii) Situation appraisal:
      (a) What is the nature of the event or the area of operations?
      (b) What is the situation or attitude of the opposing side?
      (c) What is our own situation?
      (d) How do our forces compare with theirs?
(iii) The comparative capacities of both sides:
      (a) The advantages and disadvantages of our mode of operations and the prospects for implementation.
      (b) Determining the most suitable form of operations.
(iv) The decision.
(v)   Carrying out the decision:
      (a) Who will carry it out?
      (b) When?
      (c) How?
      (d) Where?
      (e) What is to be done?

During this process the students are also taught another important item.

If the mission is not very clearly defined, it's up to you to look at the situation and decide what your mission is. You must not wait to be told what to do.

'To derive a mission from a situation' is a point meticulously observed by the Turkish soldier. This logical order produces stereotyped thinking, something which has entered the Turkish Army with the American system of training. It is the method of reducing an event to its simplest terms through questions that must be answered by a 'yes' or 'no'. While this method is very helpful in understanding complex topics, it leads to impossible situations and dangerous results when taken to extremes, or when people try to apply it to areas where socio-political elements and social traditions and reactions also play a role (like solving certain political or social problems). Stereotyped thinking, nevertheless, is still one of the methods most used by armies.

*   *   *

. . . in Ataturk's words, welcome to the School of Discipline . . .

The Military Academy is indeed a school of discipline. The commander-to-be is taught discipline with everything; it is the kind that will be strictly observed, and will mean executing orders without question and sticking to the regulations down to the last detail.

Discipline is strictly applied in the Military Academies to ensure that its mark remains indelibly on the commander's life after he leaves the Academy and that there should be no danger later on of slackening as a result of practical difficulties in the field. Every student is credited with 160 disciplinary points when he enters the Academy. The points are like gold dust. At the end of the year, his card is credited with his remaining disciplinary points; this report card is the most important factor in determining his promotions all the way up to the rank of general. He will lose points if kit inspection reveals any infringement of regulations, if he has been seen fooling about in the grounds, or reported for sloppy dress or comportment while in uniform in public, with hands in pockets or hat on the back of his head, and so on and so forth – there is no end to the regulations.

The standards of discipline of the Military Academies apply equally to youths from civilian and Military Schools. At first it is hard even for students from the Military Schools to adjust to this level of discipline because the standard imposed by the Military Academies is that much higher. Those in charge of

training consider high standards essential in order to impart
to young men from civilian schools the amount of knowledge,
the new attitudes and modes of behaviour they must absorb
to divorce them from their civilian environment and turn them
into commanders.

> If we are not strict here, if we don't implant discipline in his head,
> who is to do this for him after he joins a unit? We have to be hard
> on him here so that he can inculcate these qualities in the men in
> his unit.

One commander put it this way:

> Our approach to discipline is, to some extent, the result of
> attitudes ingrained in the Turkish people. If you adopt a soft
> approach, that triggers a tendency to walk all over you. If you
> attempt to show a bit of understanding, it is seen as a sign of
> weakness and brings familiarity. And familiarity is the one thing
> the military profession can least afford.

Another commander explained that the social structure of
Turkey and the current economic climate have both influenced
discipline.

> Our weapon systems are old and obsolete. Modernization is
> under way but it takes money and we have still a long way
> to go. We have to make do by having more men under arms.
> Put these two things together and you will see that discipline
> is doubly important for us compared with the Europeans or
> Americans.

When one mentions soldiering, discipline comes first to
mind. There cannot be soldiering without discipline. When a
commander has to send out a man to face death, his strongest
tool of persuasion is troops conditioned to obey and carry
out orders without question. This matter is taken seriously,
particularly in Turkey. According to military instructors, 'the
Turkish individual will not perform a given task without
discipline; perfect performance therefore depends on good
discipline'.

So, the most important characteristic of the Turkish Army is an extremely tough brand of centrally organized discipline. The military instructor who said 'if every officer took to debating an order issued by a superior, no army in the world could function' may have been right, but doesn't the Turkish concept and practice of discipline (due perhaps to the make-up of Turkish society) destroy desirable qualities like initiative?

> I never argue beyond a certain point with an officer of higher rank than myself. I've tried and been told off. It's regarded as bad form and you are immediately labelled a trouble-maker and a know-all. Some fellow officers do tend to be pushy. If you're quite brilliant, you may get away with it for a bit; but if you're an officer of average ability and you speak up once too often, you end up the loser and very quickly too. It's safest, therefore, to avoid any fuss.

It is possible that the lieutenant who said this may have exaggerated. In fact, a commander with long years of service in the highest ranks of the army clearly disagrees with such views:

> On the contrary, the unit commander encourages his men to discuss their ideas as widely as possible during training. There are limits, of course. Once the commander has announced his final view and decision, the debate is over and the executive phase begins. Responsibility rests with the Commander. It is he who makes the decision, and it is he who has to face the consequences. The extreme discipline practised at this point will influence the entire life of some of our young men. Some take it to heart to such an extent that they begin to regard laughter at a moment of joy, or tears in a period of great anguish, as a sign of weakness. They are unable to relax, and walk on eggshells all their lives. They may be in the minority, and it is everybody's wish that the kind of discipline we apply should be less for show and more of a 'real work discipline'.

As the country develops and its economic standards rise, with a corresponding rise in educational standards and the emergence of new values, the practice of enforced discipline will automatically diminish. This has already begun to happen in certain areas and posts.

Regulations enter as a factor in the life of the budding commander, remaining and proliferating for the rest of his professional life.

There is a written regulation for every single thing, from the simplest to the most complex, setting out what should be done, where and how. Everything, from keeping a toilet clean and turning off the light when leaving, to the notice on the Commander's office door which states what he will be doing at any given time. This affliction, this red tape, exists in every army in the world. But the Turks, probably due to causes inherent in Turkish society, have an absolute plague of regulations, from how to change an electric plug, to when and how to service a gun. There is another practice which also probably stems from national temperament, namely drawing up regulations and then ignoring them. Turkey shall have a healthier society the day 'we do what we set out to do, and avoid setting out to do what we can't do', as one Commander put it.

This tradition of regulations is both the strength and the weakness of armies; ingrained habits of dependence on rules and regulations blunt the capacity to respond flexibly and to make rapid adaptations to actual conditions.

> The military profession cannot tolerate errors; it can only operate with machine-like regularity through categorical regulations, leaving nothing to chance. This is the only way to minimize faults in the machinery and for the wheels to turn, slowly, but with a certain regularity.
>
> It is possible that an overzealous application of this mania for regulations is responsible for the saying, 'the military are inflexible'.

4

'You are an Officer. You are Superior'

The heaviest burden the budding commander at the Military Academy has to bear is the sense of moral responsibility. In addition to study, the young men are subjected to particular moral themes which are carefully developed in addresses by the Commanders and in speeches and conversations on certain anniversaries.

The underlying purpose is to emphasize in various ways 'how different they are from those outside'. It is clear from the beginning that the cadet will not achieve much material affluence even when he gets his foot on the promotion ladder through his first posting. It is clear what the young man will earn in the years ahead. To compensate for the fact that soldiering is not a profession leading to a brilliant material future, its moral aspects are emphasized instead. Whether they speak in meetings or classrooms, the Commanders nearly always elaborate on certain themes:

> Lads, you are dedicated to a duty that is too important to be measured in material terms. There is not enough money in the world to buy one's way into the realm you have entered. You all know that money cannot buy happiness. You are men with the chivalrous spirit of the knights of old. This is a profession of self-sacrifice and you are dedicated to our glorious banner and the motherland, expecting nothing in return. In this respect, you are also extremely fortunate as there are millions of youths out there who would give anything to be in your shoes.

On ceremonial occasions, these speeches are so effective that some of the students faint with excitement.

43

Then comes the eulogy of the cadet's superiority.

> Always bear in mind that you are superior to everyone and
> everything and that you are trained here to have superior
> knowledge and superior qualities. You have dedicated your
> life to the country without reservations, you are selfless and
> honest. As officers of the army which has inscribed the most
> glorious pages of Turkish history, you are different from your
> contemporaries abroad, and from other officers elsewhere in the
> world.

These speeches run to great lengths, limited only by the
eloquence and imagination of the Commander. Lavish efforts
are made to teach the young students what cannot be taught in
normal lessons. The aim is to ensure the total dedication of the
young cadets to the army, and to get them to understand fully
how important and glorious is their task.

This 'special training', which begins at school and continues to
the end of the Military Academy, proves extremely effective for
the great majority. By the time he is in the graduating class, the
aspiring officer has begun to see himself as genuinely different.
He develops a sincere and intense devotion to the army and his
fellow cadets. This is the kind of officer that the Turkish Armed
Forces are trying to create. Does the young man's enthusiasm
remain at the same level when he leaves the Academy and comes
into contact with the real world on his first posting, or do these
sentiments fade in the process?

As we shall see, there are naturally some disappointments.
But after his 'intensive' processing at the Academy the Turkish
officer is left with many more concepts like 'devotion to the
motherland', 'self-sacrifice', and 'superiority' than those who
have not been subjected to a process of similar intensity.

Another point persistently and strongly dwelt on is the
question of the appearance presented to the outside world.
Admittedly, we all care about 'what people think of us', but
in the Military Academies this is a matter of special concern.

> My boy, the uniform symbolizes not only your own honour but
> the honour of the army. A missing button on one's uniform means
> a missing button for the establishment that is your whole world.

You must therefore pay great attention to your appearance. Your clothes must be spotless, your shoes immaculate.

Behaviour in the streets is also taken seriously.

Never be seen racing like a madman. Do not jump on buses and hang from the doors. Behave in a manner that befits the honour of a student of a Military Academy, and remember what people will think. Do you enjoy watching a student racing in a hot sweat and with his hat awry?

Words are not always adequate and their message is bolstered by films shot in secret that show somebody running, with buttons undone and about to part company with his hat. The point gets across. The importance attached to appearance is intended to impress the public with the image of the soldier as 'different'. It is meant to ensure that, among the dishevelled and unshaven who throng the streets, the soldier stands out clean and crisp, even striking, in his dress. It is a carefully balanced effort to reinforce the army's prestige through its outward image.

Regulations sometimes play funny tricks, of course. For instance, even if the temperature soars to the nineties, the military student who ventures out must wear his gloves, until the summer uniform is authorized. In addition, he must display outside the Academy the same discipline he observes inside. He has to conform to the overall goals set by the General Staff that 'a neat, correct officer must be clean-shaven, wear his hair short, and be mindful of all he does; from the way he walks to the way he talks'.

Lads, don't forget for even a moment that every single step you take involves the honour of the glorious Turkish army. Walk tall, chest out. Show that you are men prepared to perform a Turk's loftiest duty. Hold your heads high. If you ever lose your bearings in a moment of excitement, stop and think, asking yourself how you appear to others at that moment. Do not talk in a loud voice. In particular, never engage in horseplay in public, never trip up your fellow-students, and don't play impromptu ball-games while in uniform.

This is how important the uniform is for the officer. It amounts to an open identity card. He knows that he will be instantly noted, even when he walks in the street. Moreover, he must keep on his uniform even on days off. He has to wear his uniform and don his hat even when he returns to the bosom of his family at weekends. When he goes to a football-match, he cannot raise his voice in protest against an unfair penalty that costs his favourite club the championship minutes before the end of the game. He has to curb his reactions.

In a nutshell, the uniform looms very large in an officer's life. You might say he lives and dies with it.

Efforts are made to instill yet another quality into the budding officer, the ability to co-exist in a regulation-ridden world and to lead a life of self-denial. This is the quality of being skilled in social graces. Efforts are made to take these youths (most of whom originate far from cities like Istanbul, Ankara and Izmir, and some of whom have never seen anything larger than their own small town) and teach them everything they will come up against in their social life.

> Never forget that so sublime is your vocation that one day you may be sharing your iron rations with the simplest soldier in the field and, the next, you may be dining at a king's table. You must know the proper way to act in either case.

Not only did the Commander repeat this exhortation regularly, he also gave a practical demonstration once a month. He would invite a group of students to cocktails and dinner and explain things to them.

> You must now imagine that my wife and I have invited you and your wives to this party. I stand by the door to welcome you and exchange a few words, taking care not to take too long as there are others coming in after you. Once this is over, I circulate among the groups and have a somewhat longer chat with each one. You, in turn, do not stick together. You form groups of four or five, depending on the number of guests. Now and then you move from one group to another, chat with those you know and meet those you have not met before. In this type of party, one does not form large groups to chat to one's own friends. One

circulates. In general, you should introduce interesting topics outside your daily business. You may talk business if the host asks a question or if there is something you absolutely need to find out, but even then you must wait for an opportune opening. You must also remember that if you don't want to face questions you can't answer, or which you don't wish to discuss, you shouldn't ask too many questions yourself. You will notice that the lower half of the glass you have in your hands is covered by a paper napkin. This is to prevent condensation dripping on to the floor and to mask the amount of drink you have consumed. Anyway, you shouldn't drink too much. You mustn't mistake a cocktail party for a drinking bout with your fellows at a tavern. Now the doors to the dining room are open. Places are shown on the miniature seating-plan by the door. You make a mental note of your own seat and move over to it but you don't take your seat until your host has taken his. Note that the person I have seated on my right at the table is either the highest-ranking person present or his guest. The food is always served from the left. Don't pile too much food on to the plate. Refusing second helpings always creates a more favourable impression. Eat with the proper forks, and eat quietly. If you're the host, you lead the conversation, introducing interesting subjects. If you're a guest, you do not butt into the conversation but wait for your turn and, when you speak, take care you don't raise your voice above the others.

To tell the truth, I too learned quite a lot from this practical lesson which lasted two or three hours.

This is the way the students are taught not only how to behave at dinners or parties but also everything they need to know about social behaviour. They are taught how to dance and how to conduct a conversation that will make a good impression. The goal is to transform the student into a modern and complete man by the time he graduates as an officer, to ensure that he is as well-informed as a university graduate, if not better, and that his social manners are up to standard. This is based on the belief that any officer who behaves badly is letting down the Armed Forces.

The social difficulties and problems of the future commanders who study at the Military Academies are not very different from those in equivalent civilian boarding-schools, but vary according to the regional and family background of the young men involved. A proportion of the problems are common to Turkish universities. However, Military Academy conditions and the special nature of military training naturally contribute particular problems. While those who enter the Military Academies from the Military Schools generally adapt more quickly, difficulties are more common among those who come from civilian schools, particularly those who come from outside the big cities like Istanbul, Ankara and Izmir, and from families of below-average income. The children who come from major cities, and from families with a higher income and educational level, find it easier to tackle most of these problems. In the case of others who come from Anatolia and from families of modest means the shock is greater.

According to experienced commanders with long service in the Military Academies, one of the shortcomings of the current system lies in its inability to provide its charges with the warmth and affection of a family home. The extremely strict discipline and sustained level of mental and physical effort from reveille to lights-out can induce stress. The officials note that Military Academies in other parts of the world face the same problems. But they point out that in Western countries in particular there are special leaves and programmes to see that the students do not miss out on the 'warmth and affection' that the family atmosphere provides, and are not prevented from entering their chosen environment on their weekend leave. This said, these officials add regretfully: 'I am unable to provide an extramural life for these children.'

The main grumble of the boarders from Anatolia whose families and relations are out of reach for their weekend passes is that they are 'unable to join a circle outside the school'. The regular opinion polls conducted at the Military Academies often yield views like the following:

> I have no desire to go out at the weekends as I have no place to go and no friends to visit. A few of us classmates band together to go for walks and then come straight back to the school.

> On Saturdays and Sundays we have nowhere to go except football-matches, and even there we have to be on our guard.

We can't let all our excitement show but have to maintain the dignity of our uniform. A few of us get together and go for walks in a group in Ankara. It is difficult to meet and make friends with people. A few manage to do so but not all. The local lads and the boys from Istanbul easily form groups but don't seem very keen to include us, since they don't ask us to join them. We have not been to any parties or such like to speak of. In four years at the Academy, I have done little more than go out to tea four or five times.

The students' isolated social life is one of the most worrying concerns of the school authorities. Another problem is that of 'girlfriends' for these youths in their prime. To understand a cadet's difficulties in his 'relations with girls', one must first consider the fact that relations between boys and girls in Turkey are not easy or based on a comfortable set of standards. Add to this the student's lack of free time because of his heavy workload, the restrictions instilled in him concerning standards of comportment, and finally the additional restrictions created by his uniform, then you can begin to understand better the extent of the cadet's difficulties.

The answers we've been receiving from the latest questionnaires at the school indicate that the proportion of students who have girlfriends is steadily decreasing. Or rather, the students complain that they have less and less time to think about girls. So I have ordered each class to give tea-parties for outside guests.

This sensitive approach by the Commander is all very well, but is it enough to solve the problem?

It is not enough, of course, but what else could I possibly do?

I made a point of watching the students on Saturdays and Sundays at the Army Academy in Ankara, the Air Force Academy in Istanbul, and the Naval Academy's new premises at Tuzla and found that while the boys from the city were able to adapt to the outside world and create their own social circle, the others generally found themselves with no choice but to go out in all-male groups.

> After leaving school we've too little time to find girls. It would be easier if we were in with a group. To chase girls in the streets is really not the thing to do after what we've been taught at school. I can't get used to the idea of chasing girls while I'm in uniform, but with a bit of luck you might find a girl.

The alternative method of finding female company in the city is to visit discotheques, but that is not so easy either. The young man who is committed to wearing his uniform feels himself out of place in a discotheque.

> Everybody starts looking at me. Some of my fellow students do go but I find it undignified, and besides it's a very expensive thing to do. We get 3,000 to 5,000 TL a month, depending on our years at the Academy. Those who get an allowance from home do alright, but it's difficult to take a girlfriend anywhere if you don't, so we confine ourselves to visiting a cakeshop or a tearoom.
> A visit to a brothel is also out of the question. I can't take my uniform off and I'd rather die than go in uniform. It's not unknown for some commanders to turn an occasional blind eye to students donning mufti for these little adventures but that is still far from a solution.

The young student indeed wears a shirt of steel – a combination of the heavy workload imposed by the Military Academy, his special training and the restrictions inseparable from his uniform. Some suffer, but others actually enjoy wearing it. The uniform is the symbol that reveals the wearer as a man apart from the world outside. It distinguishes him from the millions, and raises him above them. For some (specially men from the navy and air force) the uniform helps to attract girlfriends, but for those who take up soldiering as a 'profession', the uniform is a restriction; for those who follow soldiering as 'a way of life', the uniform attracts respect from the Turkish public and confers superiority on the wearer. The Academies of all three services shape the mentality, lifestyle and world outlook of the young student for years ahead. After a while, he identifies with his uniform and cannot be parted from it.

Whichever way one looks at it and wherever one is in the world, the uniform is an important matter for any soldier; but for the Turkish officer it has an altogether special significance. It is his distinguishing symbol, a banner that proclaims his status without any word or introduction from him; it shows he is not an ordinary man in the street; it enables his word to be heard, and even, when he is above a certain rank, causes people to rise when he enters a gathering. As a result of all this, he is very careful not to disgrace his uniform and hates anyone who does so.

5
The Principles of Ataturk

Place: The Army Academy – Ankara.
Date: 13 March.

'Roll call!'
'Fall in!'
'. . . Mustafa Kemal Ataturk!'
Cadets (in unison): 'PRESENT!'

This ceremony is held in the Army Academy every year on 13 March to commemorate the day that Ataturk enrolled in 1899. Nothing is more symbolic than the cadets' response when Ataturk's name and number are called out: when they say 'present', they are speaking sincerely. The future commanders are instructed in such a way that they adopt Ataturk once and for all as their own, and gradually begin to identify with him.

Equally significant is the ceremony held every year on 27 December to commemorate Ataturk's arrival in Ankara during the War of Independence. On this occasion the cadets in the Army Academy race to the building where the Grand National Assembly was first held, wearing battle-dress to 'represent the arrival of Ataturk'. The underlying point for all to see is that the 'Ataturks are not dead but living and, if need be, will save the country from disaster, keep the Republic alive and deliver it to the nation'.

Ataturk means everything to an officer: banner, guide, expert in military tactics and helper in solving problems. It would not be an exaggeration to say that cadet-officers hardly spend an hour without mentioning his name. In a gradual process they will internalize his role as the 'saviour of our land and country', though some will not go as far as to identify with him.

From their first days in the Academy they begin to discover aspects of Ataturk which were previously unknown to them. They realize that what they had already learnt about him at school was a mere outline. Ataturk begins to take on a different image as they attend courses and lectures.

> Those who have spent years studying this subject claim that a good leader and commander must not only be intelligent and have foresight, initiative, and the ability to take decisions, but must set an example through his integrity, courage, trustworthiness, sense of fairness, and understanding.

In the course of such a lecture commanders of various nationalities are taken as examples to provide an overall picture of the different qualities of leadership represented by each. 'But in fact isn't it pointless to engage in a long search for different examples of each characteristic? Here is Ataturk who combines all qualities in his person.' Then the lecture is followed by examples showing how the qualities mentioned above applied to Ataturk.

This type of instruction, which continues throughout the years at school, is not confined to textbooks. From the simplest ceremony to the full-scale lecture, from assembly halls to meeting rooms in distant barracks, mess-halls and special 'Ataturk corners', each provides an occasion or space for the dissemination of Ataturkism. Much effort is put into instructing the future officer that Ataturk should serve as his flag, and the ultimate goal is finally achieved: Mustafa Kemal Ataturk does indeed get under his skin, becoming a leader immortalized in his mind.

Ataturkism has been especially important since 1980. A study of textbooks and interviews held with graduates of the Military Academies from 1940 to 1970 shows that this movement developed over the years.

> It did not have a prominent place in the curriculum of the Military Academies in the 1940s. There was no such thing as 'Ataturkism', because there was no need for it. It was natural, like eating and drinking. The officers regarded Ataturk as one of themselves and were at one with his principles and reforms. There was no

need for instruction in depth. Besides, Ismet Pasha (Inonu), Ataturk's closest associate, was the head of state, and there were no counter-movements to Ataturkism. We did occasionally attend lectures or were assigned to do research on the subject, but that kind of instruction did not reach the dimensions it has today.

Such is the view of officers who held administrative positions in the Military Academies in the 1940s, who were also keen to point out how the trend had begun to change since the 1950s. Indeed from that decade, Ataturkism became more prominent in textbooks and curricula, for example in the 1960s there was a rise from 5 to 8-9 per cent in the number of hours of instruction in the subject.

Discontent and unrest in the army first began when the Democrat Party was in power (1950–1960). Officers, who had so far been taught only to obey, were startled by the rise of religious reactionary movements. Of course this was not the real reason for the agitation in the army, but the officers felt that the Democrat Party was exploiting religion for political ends and making concessions on the reforms of Ataturk to attract more votes . . . After the military intervention of 1960, more time was devoted to study of the reforms . . . With the surge in left-wing activism, which brought about the military intervention of 12 March (1971) and continued throughout the 1970s, more impetus was given to the teaching of Ataturkism. However, it was after 12 September 1980 that a scholarly stance was taken in the teaching of the subject which assumed the dimensions of an ideology. The army had never wavered in their loyalty to Ataturk but, as I said, they felt the need for a more comprehensive education . . .

Currently the total number of hours of instruction in the Military Academies amounts to 960 per annum. Of these, 160 hours (i.e. about 20 per cent) are related directly or indirectly to the study of the principles and reforms of Ataturk or of Ataturkism in general. Courses, of 32 hours per annum each, in Ataturkism, Leadership, the History of the Turkish Revolution, Political History, the Law of the Armed Services and Management and Administration, constitute the basic education of Turkish officers. The events leading up to the intervention of 12 September

1980 lie at the root of the change in the curriculum. The growth of right- and left-wing movements which were reflected also in the Military Academies prompted a top-level decision to transform Ataturkism into an ideology.

Just before 12 September, it dawned on us that, caught as we were in the tug of war between left and right, we had to develop a firm ideology of our own. It was this need that gave birth to the idea that we had to give more weight to Ataturkism . . . But this turned out to be a big problem because we couldn't find an academic outside the radical left or the extreme right capable of writing a textbook on Ataturk. As we weren't pleased with the books already available, we got down to work ourselves. General Oztorun initiated an extensive study compiled in three books to enable us to move on to a systematic form of education . . .

Is Ataturkism an ideology? It is indeed, according to one of the generals who has contributed most to its dissemination in the Armed Forces:

Communism is considerd an ideology but in my opinion it's Ataturkism that really fits the definition of an ideology, as it has an answer for any event or development. It is dynamic, not static like communism. We put this ideology into practice in every field, in the family as well as in education.

The 'principal command' which serves as a preface to the textbook prepared for the course on Military Leadership and Ataturk reads as follows:

1. This book, compiled by the General Staff as the first of three volumes to be studied in the military schools, aims to unite the commissioned and non-commissioned officers of the future through the principles of our very own Ataturkism so that they can systematically cultivate Ataturkism and absorb it as an ideology. It also aims to give a comprehensive explanation of Ataturkism, covering all human activity.
2. The first book consists of those manuscripts of Ataturk and his oral statements pronounced on various occasions, which are

directed to the future and have lasting value, thus reflecting only the views and directives (ideology) of Ataturk . . .

Kenan Evren

General
Chief of General Staff

General Oztorun's three volumes mentioned above cover almost every issue that was of interest for Ataturk, from the principles that he wanted to put into practice in Turkey and the necessary course of action to reach his objectives, to freedom of the press, trade unions, and how political parties and democracy should work.

The major point of difference between education in the military and the civilian schools is the emphasis given in the former to a serious and highly detailed grounding in history. 'Turkish History', which is not taken seriously in civilian high schools, is one of the most important courses in Military Schools. It covers in great detail the period from the battles of Kosovo (1389) and Mohacz (1526) to the War of Independence and the foundation of the Turkish Republic. The causes for the rise and decline of the Ottoman Empire are also the subject of detailed analysis.

The decline of the Empire and the War of Independence receive particular attention. Major points of interest include the erosion and weakening of the Empire by foreign powers and the 'capitulations', and the betrayal of their homeland by the Ottoman rulers. Court intrigues, political feuds, and the promotion of self-interest to the detriment of patriotism are studied in depth. In the study of the rise of the Ottoman Empire from the unification of the Turkish states to its decline and fall, particular emphasis is given to the lack of patriotism and nationalism on the part of the administrators. While others factors are also taken into account, the main point of the argument is that 'no-one stood up for the homeland'.

The study of history in the Military Schools is not confined to lessons. Library loans show that books on history, geography and literature constitute 54 per cent of those borrowed by the students. The majority of books on loan are those on Turkish history and geography in particular, and in literature works on heroic exploits are generally preferred. But cadets do not do much reading outside their coursework and are not particularly interested in world events.

Cadets, already involved in extensive historical study, receive an even deeper grounding in how Ataturk saved the homeland. The War of Independence is taught to the future officer in such a way that he can follow almost every moment in Ataturk's course of action and in the pattern of daily events. He learns almost every battle by heart, including the tactics and politics it involved. It can be said that he shares vicariously in the experience of rescuing the land from the ruins of the Ottoman Empire, and founding the new and modern Turkish Republic with Ataturk.

The War of Independence is taught so systematically and with such enthusiasm that the cadet internalizes it with an emotional involvement unknown to his civilian peers. Through repeated ceremonies, classes, lectures and performances, his mind absorbs the War, and the huge transformation his country underwent in its salvation. The names inscribed at the entrance to the school of those who fell fighting in the War serve as a daily reminder.

This type of historical grounding in school, or, to be more precise, vicarious experience of the War of Independence, is the key to a sound understanding of the officer's attitude once he is out of school and rises in rank. It is also the key to his views on democracy, political parties, and ideologies. It stands as the period when he identified himself with Ataturk and gradually adopted the idea of 'saving the homeland'. Having carefully studied Ataturk's rebellion against the politicians which started the revolution and destroyed the existing order, and the methods he used to rally the Anatolian people, the officer believes that 'should there be a need, such a course of action is legitimate and even necessary'. Naturally the military aspect of Ataturk is emphasized rather than the civilian; for example, Ataturk is generally dressed in a field marshal's uniform in the pictures decorating military premises. After four years of instruction, the officer looks upon Ataturk, the man who has stood up for his country and raised it from the ashes, as his guide and leader.

Education in the military schools is highly systematic and very carefully planned. The following words of a top-ranking member of the General Staff sum up the objectives of the military educational system:

> An officer's awareness of identity derives from his sense of superiority over a civilian, and from his ability to understand the finer points of patriotism, unchauvinistic nationalism, and the Ataturkist way of thinking, and also from his ability to accept that he can die for his country. Although he may not be conscious of it, this is the way his mind has been conditioned. Some officers faint with excitement while saluting the flag. The civilian does not cultivate such qualities. From the same substance we manage to craft something fine, while civilians can produce nothing comparable . . .

Brought up on the history and enthusiasm of the War of Independence in the conviction that they were at Ataturk's side as he laid the foundations for the Turkish Republic, cadets are also made responsible for 'guarding and protecting the existing order' and for keeping 'Ataturk's torch alight'. But in fact these are points that require no emphasis. Who would be expected to protect Ataturk's reforms if not those who come from the same school and hearth? The cadets take on the duty naturally and assume their responsibility as standard and torch bearers. In other words they perceive 'from the circumstances' that it is their duty and they have no trouble in adopting it as their own from the start.

However, the objective is not only to symbolize Ataturk as a torch for future officers but to instruct them in the thoughts and directives of the Great Leader, namely in his ideology. Thus cadets have to study extracts from Ataturk's speeches containing his directives in almost every area, from his ideal vision of Turkey to the economy of the country and the functioning of democracy. They begin by learning that Ataturkism is not a static ideology but one that undergoes a 'dynamic' development according to the conditions of the time. Hence it is not inflexible like other ideologies, but keeps changing in the light of previous experience and progressing towards the 'ideal'. This is emphasized as the 'Dynamic Ideal'.

The following extracts from textbooks may serve to illustrate readings in Ataturkism. In the Introduction to one of them the origin of the ideology is explained as follows:

> In his statements Ataturk points out that a 'powerful state' is the most effective means of achieving the Dynamic Ideal . . . The tasks that face the Turkish nation in achieving the Ideal will be determined and put into practice in a continuum according to circumstance, means, and potential. Ataturkism embodies the essential principles for the best ways of action that will direct the Turkish nation towards the ideal. Ataturk founded institutions that would comprehend, explain, and propagate, from one generation to the next, the ideology underlying the issues of the Turkish nation. In sum, Ataturkism is real and consists of the values, principles, and essentials which will lead the Turkish nation to success, because it embodies conclusions drawn from the multitude of disasters and sufferings recorded in the history of our homeland, our nation, and of many other nations.

The definition and major significance of Ataturkism is explained as follows:

> Ataturkism is the set of realistic ideas and principles concerning the state, the economy, intellectual life, and the fundamental social institutions. The basic principles were also laid down by Ataturk to ensure the full independence, peace, and welfare of the Turkish nation in the present and the future, to ensure the sovereignty of the nation as the basis of the state, and to raise Turkish culture to the level of modern civilization under the guidance of rational and scientific principles.
>
> The adoption of Ataturkism on an individual and nation-wide basis and its protection against current and prospective movements of a deviant or conservative nature serve as the guarantee for the development, strength, and enlightened future of the Turkish state.

Nationalism is emphasized as being primarily important in relation to the 'unity, solidarity, and integrity' of the country:

Unity and solidarity are characteristics of the Turkish nation. They signify the integrity of the nation and allow no room for divisive and separatist elements. National unity, the sign of strength and power for all nations, is our most cherished possession . . . Ataturk drew attention to the necessity for national unity and solidarity which are the major elements of Turkish nationalism, and warned against any course of action that might endanger them . . . In terms of Ataturkism, the principle of populism, which is synonymous with democracy, protects the country against separatist claims and class struggles . . .

The Ataturkist conception of republicanism, and of democracy in particular, holds an important place in the education of cadets. Their attention is drawn to the significance of the democratic form of government for the Turkish nation and its historical development.

A republican government is that which suits best the character and customs of the Turkish nation. A republican regime implies a democratic state.

Such are Ataturk's introductory statements in the following extract which explains the importance of the will of the nation, of the national assembly, and of populism in relation to both:

In terms of Ataturkism, democracy and populism mean the same thing. 'Where governments are based on democracy (populism), sovereignty belongs to the majority of the people. The principle of democracy, requiring that sovereignty should belong to none but the nation, is related to the source and legitimacy of political power, i.e. sovereignty.' Such statements emphasized Ataturk's view that democracy derives from populism . . . 'In ancient times the Turkish nation proved their adherence to the idea of democracy in the general assemblies they held to elect heads of state. But later on this principle was violated by the despotic Ottoman monarchs.' Ataturk pointed out that populism was like an ever-rising sea which found its way into all modern constitutions.

Explanations of Ataturkism include trade unionism and even associative movements among theories and ideologies such as communism which run contrary to democracy:

Ataturk, who found fault with theories opposing democracy (populism), helped to provide a comparative evaluation and definition of populism: 'This concept of democracy has come under attack by Bolshevik theory, revolutionary political syndicalism, and the theory of representation of interest groups. Let us explain why they are wrong in opposing our concept of democracy:

Under the Bolshevik theory, a minority composed of workers, and officers of the navy and army have united in the economics-based Communist Party and established a dictatorship. They have no national goals, no respect for the sovereignty of the people, and recognize neither equality nor the freedom of the individual. At home they forcibly impose their own views on the majority of the people, and abroad they try to spread their principles to the international community through propaganda and revolutionary organizations. However, the first aim in setting up a government is to guarantee the freedom of the individual. The Bolshevik form of government is arbitrary. If a society is forcibly enslaved to the opinion of only a minority and is therefore incapacitated, one cannot consider such a system of government a natural or rational one.

Theoreticians of revolutionary political syndicalism are workers' groups who would have all political organizations work only in their interest, so that they can eventually take over political power and sovereignty. Waiting for an opportunity to achieve their aims by force, they occasionally organize general strikes, thereby making their presence felt and exercising their influence on the government to have certain matters solved in their favour . . .

The theory of the representation of interest groups is rooted in the diversity of interests represented by various professional groups, artists, and businessmen, each group being a separate entity in society. Thus, it is claimed that such groups will seek their private interests and, whatever the outcome, this will depend on the extent to which such interests are promoted. Who will stop them from working only in their own interests if some of these groups join forces in the representative assembly and come

to power? Accordingly, we do not think that either this particular theory or others mentioned above are suitable for our country and nation . . . In our view, farmers, shepherds, workers, tradesmen, artists, soldiers, doctors, in short all citizens who belong to a social institution are equal in respect of rights, interests, and freedom.'
According to Ataturkism, the people can be defined as a collective who have adopted equality before the law and who attribute no privileges to any family, class, group or individual. Those who adhere to this essential principle are for the people and can therefore be regarded as populists.

In populism, the foundation of social order is directly related to the principle of work:

Ataturkism has not only adopted the principle of the equality of the Turkish people before the law, but has also defined their responsibility, which is to work. In Ataturk's view, the survival of society is endangered if the individual refrains from work. According to the principle of populism, social order in Turkey can be preserved and maintained by the labour of the individual. The following words of Ataturk clearly point to the direct relationship between populism and the principle of work: 'Let us be fully aware that we are a people who must work for salvation and survival. We all have rights and powers, but we must work for our rights. Those who are idle and shun work have no place or right in our society. Hence, populism is a social system that depends on labour and law.'

The concept of a powerful state is presented as the most striking aspect of Ataturkism. That the state must intervene wherever necessary and that such a powerful state serves as the most effective means of achieving the Dynamic Ideal is emphasized throughout.

Ataturk's etatism is based on the essential principle of individual work and activity. Nevertheless, in order to achieve the Dynamic Ideal as quickly as possible, it has adopted the view that, in the general interests of the nation, the state should be actively involved in every field, especially the economic. The active involvement of the state signifies putting into practice and implementation, and includes orientation, encouragement, aid, organization, and inspection.

Secularism is among the most extensively taught and heavily emphasized principles of Ataturk. A point underlined in this respect is the possibility that, unless state and religion remain separate, the country may once more revert to a period of decline and may have to surrender to foreign powers as a result of losing touch with developments in the modern world.

Finally, the readings on adherence to Ataturk's principle of revolution suggest the course of action for the development of the country:

'Revolution means changing the established institutions by force. It signifies destroying those institutions which hindered the progress of the Turkish nation in recent years and replacing them with new ones which will raise the nation to the highest level of civilization.' These statements by Ataturk expressed his wish for Turkish society to adapt itself to the requirements of the age, to develop and renew itself. But as the following statements show, Ataturk envisaged a swift rather than a gradual renewal: 'Is Turkey to make her progress step-by-step or with a sudden leap forward? There are two methods to consider. One is the well-known way of the French Revolution, by which regimes change, counter-revolutions follow, the right trample upon the left, and the left sweep away the right, and the next thing you know is that one hundred and fifty years have gone by . . . Does our nation have enough blood in her veins or enough time before her? . . . If I have sufficient power and authority, I think I will bring about the revolution desired in our society by means of a "coup" . . . Having spent so many years of my life educating myself, studying civilized society, and enjoying freedom, why should I stoop to the level of the people instead of raising them to my level? They should be like me, not I like them!'

Instead of the use of persuasion and education, Ataturk's revolutionary principle envisaged a short cut to achieve the means of elevating the people to the ideal standards he set for Turkish society. This way of thinking finds sympathy not only with the young cadets in the Army Academy but with many other groups in Turkey.

WHAT DO THE CADETS MAKE OF THE 'IDEOLOGY' OF ATATURK?

The six principles of Ataturkism (i.e. republican, nationalist, populist, etatist, secularist, and revolutionary) are covered extensively and in great detail, with examples from Ataturk's speeches and interpretations of them. The ideology to be cultivated by the cadets has a wide scope, ranging from the Dynamic Ideal to the spirit of solidarity.

It must be noted, however, that the cadets who are about to graduate do not seem able to grasp Ataturkism as an ideology. This may be due to a failure in the lectures and textbooks, to lapses in presenting the material convincingly and scientifically, or to their unwillingness or inability to accept Ataturkism in such a way. The interviews I held with senior cadets, for example, gave me the impression that they had adopted Ataturk more as a banner than as an ideologue.

'What does Ataturk mean to you?'

The cadet's eyes shone as he responded to the ignorance underlying the question:

'. . . Ataturk is the man who saved our country, founded the republic, set up the principles for the development of our nation. He is our leader . . .'

'Do you believe in his ideology?'

He replied without hesitation, perhaps in reaction to my use of the word 'ideology': 'We have nothing to do with any ideology. We are followers of Ataturk.'

'But you did learn about the ideology of Ataturk?'

'That's different. We were taught about the principles of Ataturk, and his way of thinking.'

A point that has also been frequently repeated by the teaching staff is that a considerable number of the young people in the Military Academies have difficulty in adapting fully to the intensive courses on the principles and ideology of Ataturk. The courses seem too abstract for them to study in the depth required by their commanding officers. No doubt there are many among them who are capable of recounting from memory what they have learned, or of keeping up a discussion, but the majority end up talking about the six arrows of the Republican People's Party symbolizing the six principles. (This is perhaps due to the notorious habit of education in general to summarize and schematize.) Neither in their school years nor later do they seem able to progress much further than a superficial interpretation of the six arrows.

The inability to internalize Ataturkism as an ideology may be due partly to the fact that it is not really an ideology, and partly to a complacency on the cadets' part that classwork, along with some additional lectures and prescribed readings on Ataturk, is enough. Although the libraries are well-stocked, they have not formed the habit of engaging in extra reading or in-depth research.

Officers accustomed to dealing with hard facts cannot identify with an abstract ideology to the same extent as they identify with Ataturk's deeds in the War of Independence. What remains in their minds at the end of their education is something both stark and simple.

'The points we are most sensitive about concern derogatory remarks about Ataturk and opposition to his fundamental principles and reforms.'

'Well, what are those principles and reforms? What would you say if I asked you to tell me the kind of change you would not be able to tolerate?'

'The first is any threat to secularism. I would not permit that . . . Then, any separatist movements that would divide the country. And communism will have to confront our firm opposition.'

The method of teaching Ataturkism may be summed up as one that focuses more on Ataturk's ideas than on the way he developed his strategy in the War of Independence and on the way his mind worked. It is evident that such a method is responsible for some question-marks and a confusion of concepts in the minds of some cadets.

The confusion becomes evident particularly over ideologies. The clearest example concerns the left. That there are significant differences between communism, socialism, and social demo- cracy, and that each is opposed to the other, despite reciprocal respect for opinions on certain basic issues, are points that are ignored. There are very few indeed who are aware that social democrats in particular, and socialists to a lesser extent, have been blamed by communists for 'adopting ideologies created by the West to undermine communism'. The dominant tendency is to regard the left as a single concept.

The principal reason for this is that liberal views, let alone social democracy and socialism, are not allowed to penetrate the Military Academies; and that contact with right-wing or left-wing ideologies is kept to a minimum. While right-wing views face as much restriction as those of the left, they nevertheless take their place under the umbrella of nationalism and do not run up against the inherent conservatism of the Academies. Thus they are not regarded as being as perverse or dangerous as left-wing ideas. It would be more correct to say that the Academies are particularly guarded about attitudes promoting a 'fascist dictatorship and theocratic state'.

> My sons, the way to conquer Turkey is to conquer the hearts of the young members of our glorious Armed Forces. Any conquest from within the army will mean a firm grip over the whole country. You must therefore be very careful on your way to maturity. If you were offered a woman, you would have to ask yourself why and be on your guard . . . Or if a pretty young girl offered you her friendship, you would have to go and take a look at yourself in the mirror and think whether you are handsome enough to deserve such attention . . .

Such warnings from their commanding officers are very clearly imprinted in the minds of the young cadets. Suspicious attitudes towards the outside world are cultivated in the same way in the Military Academies, regardless of whether the object is Turkish or foreign, civilian or military.

The fundamental purpose of the Academies is to keep the young men away from political ideologies and movements. Therefore books and periodicals stocked in the libraries generally have a conservative outlook. Anyone seen with publications not stamped as 'readable' by the head office is penalized. For instance, left-wing and religious publications are not allowed. Similarly, care is taken to invite external lecturers or guest speakers who are known to be conservative not liberal in their outlook. In this way it is hoped that cadets will have minimum access to political trends outside the ideology of Ataturk and more time to devote to their studies on Ataturkism.

One may well ask if this is the best method for achieving the desired end.

> In 1945 I was a lieutenant on the warship *Yavuz*. One thing I can't forget is that one was allowed to read nothing but the names of the cabinet ministers. In 1958 I enrolled in the Academy. One day my commanding officer, a nice man, saw me reading the *Akis*, a political weekly. 'Don't get yourself into trouble, my boy', he said, 'read it at home'.

Such memories date back only a few decades, not to the distant past. Nowadays Turkish officers are no doubt trained to have a much broader outlook on the world and are better read. Nevertheless, one wonders if the conservatism currently imposed on them does not provoke some reaction or arouse their curiosity. Would it endanger the future of a young lieutenant or would it make him a wiser officer if he were allowed to find out what communism is and what it is not?

There is also a confusion of concepts regarding economic matters. The Ataturkist model for a mixed economy, encouraging the state to intervene, orientate, and motivate economic life, inevitably appears to contradict the model for a liberal economy implemented in recent years. The confusion spreads as more questions are raised on the subject: which is the better, the Ataturkist model or the one promoted by Prime Minister Ozal?

The former is known to have been initiated in the 1930s when it was felt that, for want of a private sector which had not yet developed, the state had to boost the weak economy. While the fundamental anxieties underlying the Ataturkist model are recognized, the picture is still not clear in the minds of many. As no further study is undertaken in search of a synthesis either during the school years or after, the issues remain obscure, thus reflecting a confusion of concepts that is to be observed in almost every section of Turkish society.

6
Politics and Political Parties

A major rule observed in the Military Academies, in classes and lectures as well as in informal talks and discussions, is that officers should avoid political involvement. Future officers are constantly reminded of armies which have deteriorated as a result of internal disputes brought about by involvement in politics.

The following saying of Ataturk is one that every cadet in the Military Academy knows by heart:

> Gentlemen, commanders should keep their minds free from the influence of political concerns when they consider or perform their martial duties . . . The principle of isolating the army from politics is fundamental to the Republic.

However, in the same speech Ataturk goes on to assert that 'in pursuing this principle so far, the armies of the Republic have remained in a position of respect and power as the strong and reliable protectors of the homeland'.

Ataturk's conception of the army as 'the guardian', in the general context of his speech, leads to the conclusion that 'the officers' political non-involvement' actually means personal detachment from party politics rather than restraint from political intervention. This is one of the reasons why in the three previous cases of intervention – in 1960, 1971 and 1980 – the army felt that it should return to barracks as soon as possible, hold elections, and leave government to the civilians. No doubt, there were other reasons too for this course of action. Nevertheless, in the eyes of an officer 'politics' is dirty work, since it tends to favour personal interests rather than speaking out the truth in the interests of the nation. Thus an officer who is politically minded or actively engaged in party politics is not looked upon favourably by his colleagues, even if he is retired: he would be condemned in such terms as 'pity about our commander!', or 'he wasn't the sort to get into that mess'. If an officer is thought to be deliberately vague about revealing his intentions, he is accused of 'talking like a politician'. The tendency, especially in the lower ranks, is to speak one's mind or keep silent if in doubt about giving offence, as this is the principle they have been taught to observe. But as they rise in rank and take on more responsibility, some are seen to modify their attitudes towards openness of speech. A feeling that becomes more prominent as they move up in rank is that critical statements that might displease could affect their future.

Although cadets in the Military Academy learn to cultivate a dislike for politics, their courses on the principles of Ataturk provide them with extensive grounding on what political parties and politicians ought to be like. For instance, the fundamental principles for parties to bear in mind are particularly emphasized in a chapter of the three-volume text on Ataturkism, beginning with the following statement: 'parties are responsible for achieving the Dynamic Ideal of the state'.

The same point is brought up in a Commander's speech:

> Ataturkism is the set of realistic ideas and guidelines concerning the state, intellectual and economic life, and the fundamental social institutions, whose basic principles have been set down by Ataturk. As this definition indicates, Ataturkism is a system of thought, an ideology which is universal and which may be applied in any country. So far as I am concerned it can also be defined as a belief . . . It rejects capitalism and socialism, and aims to achieve the Dynamic Ideal by eliminating religious discrimination and influences of a divisive and separatist nature . . .

The Commander goes on to stress the overall aim in the following terms:

> Ataturkism explains the duties and courses of action for those in responsible positions in the fundamental institutions. If Ataturkism can be properly interpreted by persons in responsible

positions with special regard to their own functions and circumstances in diverse fields of activity, then the duties of each
individual in Turkey will become much clearer . . . To grasp
the objective in Ataturkism is to understand Ataturk. Secondly,
as we define our duties, perform them, and create new ones, we
must determine the means and possibilities for achieving them.
Thirdly, we must plan in detail the course of action that will allow
us to perform our duties successfully . . .

Following this general outline, cadets in the Military Academy
are instructed in the principles of republican democracy: what
the political parties representing the voice of the people in the
Grand National Assembly should be like, what qualifications
politicians should have, and how the opposition should function.

Political parties are responsible for achieving the Dynamic Ideal
of the state (i.e. the Dynamic Ideal of Ataturkism) . . .
Observing this objective constitutes the basic rule underlying
the programme of each party. The aim of 'establishing and
administering a firm state authority which will protect all the
reforms, ensure the full security of its citizens and order and
discipline by means of internal and judicial organizational laws'
must form the basis for all activities of the state and its institutions
in the Republican, Nationalist, Populist, Etatist, Secular, and
Revolutionary Turkish state . . . Parties participating in the
administration of the state as representatives of the whole nation
must be realistic and envisage meeting the needs of the country as
well as achieving the Dynamic Ideal. Real democracy for Turkey
lies in the principles of Ataturk and especially in the principle
of populism. Thus democracy can achieve its real purpose only
through elections which faithfully represent the views of the
people . . .

Such statements emphasize that political parties should
remain within the principles and all-inclusive views on
Ataturkism which are taught in the courses.
The cadets are also taught how parties should function:

Political parties should not promise to meet all the wishes of
the people. This is damaging for democracy, and fails to be

convincing. It is natural for the people to press for their needs, but they cannot fully consider how this can be accomplished. Political parties, on the other hand, have to consider the people's needs and wishes in the light of the existing means and of the future and general well-being of the country. The ideal of the Turkish state should be the Dynamic Ideal of the parties; the essential principles underlying the state should be adopted by all the parties and Ataturkism should serve as the common ground for all . . .

The practice of continuing support for anti-Ataturkist trends or banned views by adopting different names for the parties is found objectionable:

According to Ataturkist principles, the party programme is the major element to be taken into account in political parties set up under various names by various people. A change in the name of the party should not be intended to 'deceive'. This point, which reflects a certain amount of irritation, is illustrated by examples of the Justice Party and the Welfare Party as successors to the Democrat Party and the National Salvation Party respectively.

Cadets are also instructed on how the opposition should function:

According to Ataturkism, parties and individuals in opposition should try to disseminate their own political views and be critical of the party in power, but their criticism should not be damaging to national unity. They should be constructive, rational, and realistic in their opposition, and bear in mind the interests of the country in criticizing the mistakes of the party in power . . .

While it is easy to understand what is meant by 'damaging national unity', the same cannot be said for 'constructive opposition'. Thus the interpretation of this concept depends very much on the personal views and tendencies of individuals.

The raison d'être of political parties is not to divide the nation or to incite clashes of interest or class but to promote the development and prosperity of the country while abiding by

the principles of national independence and sovereignty . . .
According to Ataturkism, the struggle between the parties should
be of a positive and constructive nature. There should be no place
for divisive politics which would run against national interests. In
Ataturkism political parties are conceived of as a bridge between
government and the people to communicate the demands of the
people and ensure that government resources are utilized in the
interests of society . . .

Such prescriptive statements about political parties in the
context of what appears in the press and the attitudes of some
politicians do not have a positive impact on the future officers,
who tend to think: 'Our parties do engage in destructive politics,
can be divisive and create hostility among the people, and do
make promises which they cannot keep.' From the start, the
image of political parties is a doubtful one, reflecting a lack of
trust on the part of the cadets.

A high-ranking general points out that this is perfectly normal:

> Is it just the cadets who look upon politicians with distrust? Don't
> the public have the same idea when they read about the deputies
> calling each other names in the National Assembly? Why is all
> adverse opinion attributed to our cadets? If this is the consensus
> of opinion, it's not their fault, is it?

Another issue that bewilders young minds concerns relations
between the government and the opposition. It is common
practice in all democracies for the opposition to challenge the
government and take it to task. Opposition may be voiced in
street protests or mass meetings backed by the trade unions.
The history of democracies is full of harsh conflict and even
bloody struggles, but eventually certain democratic principles
provide a common ground for agreement. What seems to be
misunderstood in Turkish society is that the source of the tension
is not a disregard for national interests on the part of the political
parties struggling for power, but the fact that the society is still in
transition and not yet fully accustomed to the democratic rules of
the game.

In other words, the question-marks in the minds of the young
people in the Military Academies are generated by prescriptive

statements regarding politics instead of descriptive ones that explain the current state of affairs. To what extent those prescriptive statements are realistic is yet another question. Don't they seem somewhat utopian with respect to the practice of democracy in Turkey as well as in other countries of the world? How healthy is it to cultivate in young minds an image of 'harmony and co-operation between the government and the opposition'? Does this not present an artificial picture which is likely to confuse future officers when they step out into the real world?

The negative image prevails about politicians as much as political parties.

> Political parties should consist of patriotic, honest, and reliable people. The presence of insincere and insidious representatives in the Assembly would jeopardize national integrity and well-being. According to Ataturkism, party members must be chosen carefully from among those who will hold their country above everything else and not abuse their power . . . They must work for the welfare of society not for individual interests. Evil elements exist in all societies, but the nation must always be on the alert against the possibility that they might wish to promote their own interests, and make false promises work under spurious party programmes in order to gain power . . .

Self-seeking politicians are unwelcome everywhere, and no-one would disagree with views which outline what a politician should not be like. But if this is taught as part of education, is it not possible to create an overall negative impression of the majority of Turkish politicians? It is the inner values of society that decide what principles to follow: for example, to distinguish the good from the bad and penalize the self-seeking politician by not voting for him. The question of whether democratic elections are the best possible form of representation can be a matter of argument, but should it not be accepted that, despite its shortcomings, no better system than democracy is available?

Prescriptive qualities such as 'working for the welfare of the nation – being honest and reliable' are also subject to individual interpretation. There is diversity of opinion regarding the ways of achieving the welfare of the nation: for some it is the way of social democracy, for others it is full-fledged liberalism; some may believe in the 'crime of thought', others may find that the granting of some rights would strengthen the nation rather than divide it. When abstract evaluations are taught in the form of principles they lead to a confusion of concepts in the minds of young people which result in wrong assessments, an erroneous sense of mission, and an unfair image of politicians.

In the educational process attention is also drawn to the ideological inclinations of politicians:

'A great deal of harm was done by figures of heroic appearance' who set up parties on the basis of class (examples of which are to be found in other countries), who formed pressure groups based on such divisive principles as class and sect to influence national deputies, who exploited religion and traditions of the past for political purposes, who rallied round a leader and an imaginary programme to form a political party, 'who initially appeared to be friendly to the nation but, having come to power, ignored its real needs and went their own way, seeking personal loyalty, without taking heed of warnings from the authorities'. On no account should such people be allowed to assume authority or responsibility . . . It is necessary that party leaders who have a major role to play in the future of the country should possess certain abilities and qualifications, and that they should not avoid responsibility . . . Instead of encouraging class struggle their principal aim should be to establish social order and solidarity, and to harmonize different interests on the basis of feasibility . . .

THE IMAGE OF THE POLITICIAN

The Turkish officer's perception of the politician undergoes a change between his years in the Military Academy and his retirement. Although the divergence between a civilian's and an officer's image of the politician is not great, it should not be underestimated.

The civilian, being more aware of his own shortcomings, is not as sensitive as the officer to the defects of a politician and is therefore more tolerant, while the officer, conditioned by his education to take a more rigorous view and not to accept any deficiency or error, is less tolerant and respectful. For example, a lieutenant has more trust and respect for his commander than for the Prime Minister or the Minister of Foreign Affairs. The general conviction in the Armed Forces may be summed up as 'what we do, the politicians undo'. A noisy officers' meeting is generally likened to 'the National Assembly'. No doubt, politicians are largely responsible for such an image.

The root of the problem lies in the officer's dissatisfaction and annoyance with the lack of discipline and organization in civilian society. He loses the little tolerance he has when he observes such deficiencies compounded by the question of political responsibility. Because of the difference in educational background, his conception of 'the state' is very different from that of the civilian. He has a far greater respect for the state and cannot accept the politician's failure to conform to the rules taught at the Military Academy, when his primary responsibility is to protect the state.

No-one can claim that the Turkish officer's image of the politician is in general a positive one. The majority of the cadets interviewed at the Military Schools and Academies view him as one who puts his personal interests or his party's ideological interests before those of the nation. Of course, there are also politicians who have won their trust and respect, but on the whole they voice the same kind of criticism as of the general public:

A politician doesn't put his country above everything else as I do. His priority is his own re-election in four years' time, though, in the meantime, he talks a lot about patriotism. He can abuse the state by indulging in favouritism to secure his re-election. I trust very few of them . . .

Some politicians aren't half as educated as we are. Some know one-tenth of what we do. How can they govern the country? They've managed to become deputies through powerful party connections or bribes. Some of them can't even speak proper Turkish, let alone a foreign language as we do . . .

I measure a politician not by his education but by the priority he gives to country and nation. Politicians are not concerned 'for the state' as much as we are. They talk about Ataturk's principles and all that only because they're wary of us or are afraid of some reaction . . .

Some of our politicians are so ignorant that they've never really been aware of the dangers inside or outside the country.

Such opinions arise from the growing gap between the military and the civilian world. Although the essential objectives are the same, a different logic or system naturally produces opposing views.

The officer is trained to finish the job he is given, no matter what, and to carry out commands. The following incident witnessed on a visit to one of the barracks is highly typical:

The commander's jeep had broken down.

'Ahmet my boy, fix it right away.'

'Yes, sir.'

The private was back in an hour. As there was no spare part available, the jeep had been fixed with a part taken from another jeep now made useless, but the commander's orders had been carried out.

The same commander saw a burst water-pipe in the city and called up the local authorities to complain that it had stayed unrepaired for three days: 'Our men are too busy at the moment, but we'll fix it.' The fact that four days later the pipe had still not been repaired was beyond the commander's comprehension: 'Is there no-one responsible for a burst water-pipe in this country?' The commander blamed it on the politicians who were incapable of properly organizing the state.

In fact, the clash of viewpoints between the military and the politicians is a universal phenomenon. American officers, for instance, are equally eloquent about the irresponsibility and insensitivity of some of their politicians concerning the interests of their nation and the dangers that surround it. Recriminations increase particularly in times of tension.

In general, the officer who is accustomed to completing his duties without wasting time grows apart from civilian life and disregards some of its important aspects. Friction arises from a lack of dialogue between the two separate worlds which have very little in common (to be discussed below).

Although the Turkish politician is warmly disposed towards

the military, he would describe them as 'strict, inflexible and narrow-minded', but of course would not discuss any of these points in public. The American or French politician also in general views the military as blinkered and always ready to shout 'danger and war'. For instance, any discussion with the civilian and military wings of NATO or with the US State Department and the Department of Defense would reveal totally different views on each side. The military would necessarily exaggerate the dangers, while the civilians' attitude would be much milder. It should also be remembered that the military over-emphasis is largely geared to pressurizing civilian governments for an increase in funding for defence and the armed forces.

In Turkey, however, the military–civilian relationship is fundamentally different from that in Western Europe or the United States, where the supremacy of the civilian government is recognized without question, no matter how strongly it is criticized. There is an uncomfortable relationship between the Turkish military and civilians which is due to the lack as yet of a similar recognition.

The officer, taught primarily how a democracy 'should' function, discovers the real conditions in the country once he is out in barracks. His perception of politicians and political parties changes in time as he finds out more about the difficulties in the actual experience of democracy: for example, that it takes a long time for society to absorb certain issues, that it is natural for political parties to struggle for power, that it is inevitable for social classes to engage in conflict for a share of the national cake. Although the difference of opinion between a cadet in the Military Academy and a colonel is immediately noticeable and may be due to the difference between youth and experience, the tolerance that comes with age cannot be underestimated. It must be said, however, that the fundamental outlook never changes. The difference between the military and the civilian worlds must affect the views and attitudes of both. But when the officer, brought up on 'how politicians and political parties should be', discovers the inevitable conditions underlying the realities in the country, he blames the politicians rather than the conditions, and does not change his mind about this.

One of the most controversial subjects includes the relations between the Democrat Party and the Armed Forces in the 1950s, which may be summed up as a series of mutual accusations culminating in the contradictory views, disagreements and hostility that have lingered to the present day.

The textbook on Constitutional Law, used in the Army Academy, explains in detail the changes in the 1961 Constitution, thereby revealing what the army thought of the Democrat Party:

> Studies undertaken in the multi-party period displayed the shortcomings of the constitution of 1924.
>
> Political developments from 1945–60 indicated the following gaps in the constitution which could be exploited by those in favour of a totalitarian regime:
>
> – In parliamentary systems the head of state must be impartial so that, when necessary, he can act as mediator between political parties. It therefore became apparent that the status of the head of state must be accurately defined by the provisions of the constitution.
>
> – It was clear that the constitution should safeguard the political parties, the electoral system and, in particular, the by-elections.
>
> – It is also clear that the executive powers must be kept under the supervision of the judiciary and new provisions must be introduced to ensure that no measures should prevent this.
>
> The Revolution of 1960 had two objectives: (i) to reorganize the Democratic Republic according to the principles of the new constitution, and (ii) to transfer the power of government to the new national assembly . . .
>
> Our society, which had moved from the single-party to the multi-party system in an exemplary manner, failed to make democratic progress under the leadership of ambitious politicians in the period 1950–60. However, in relinquishing power after free elections, the military forces of the Revolution left no-one in any doubt that they, as dynamic forces following Ataturk's reforms on the path to Western civilization, were determined to establish democratic order and 'ensure mutual respect' among individuals . . . With the beginning of the multi-party era it became clear that the reforms of Ataturk had not yet taken a firm hold in society, and that a substantial number of votes would be cast to hinder reformist efforts. In their struggle for power, political parties capitalized on such votes and began to make concessions on the reforms. Throughout the period 1950–60 religion was

consistently and increasingly exploited for political ends and an anti-reformist attitude was adopted. It therefore became apparent that measures had to be taken in 1960 to safeguard the reforms which in 1945 had required no protection.

Ideas constitute a central, converging force in society. They may change and be renewed in time, but some are so important that, if they are subjected to change or violated at a particular moment, the result is anarchy. The Constitution of 1961 warned us that violations of Ataturk's reforms would lead to violence in society . . .

Although the Justice Party was opposed to the 1961 Constitution, they did not explicitly reject it; instead they played with words and asked their electors to vote against it. The New Turkey Party was divided and could not take a definite stand. There were differences in this respect between the New Turkey Party and the Justice Party who both wanted a share of the votes of the defunct Democrat Party. As for the Republican People's Party and the Republican Peasants' National Party, they had both taken part in the drafting of the new constitution and therefore voted in favour of it . . . The constitution was finally accepted by 61 per cent against 38 per cent of the votes.

The Armed Forces' view of politicians and their practices is made abundantly clear in the following extract on the causes underlying the military intervention of 1980:

The Armed Forces, acting with a sense of historical responsibility on behalf of the Turkish nation, executed the 12 September 1980 operation within the military hierarchy, and took complete control of the government . . . The legislative and executive bodies and other constitutional institutions failed to take any measures against the anarchy, terrorism, and separatism running rife in the country for many months; no steps were taken against the violation of law and order, no solution was found to the economic crisis; the legislative bodies remained indifferent to the nightmare that reigned over the country and chose to turn a blind eye even to the violation of the constitution; moves to elect the President of the Republic as of 22 March 1980 were taken lightly and reached deadlock on account of the pursuit of political interests, as a result of which, time, which would have been invaluable for fighting the crisis, was heedlessly wasted, 5,241 people died, and 14,000 were wounded . . . It became necessary for the Turkish Armed Forces to take over the government because, despite the heavy loss of life and the large number of wounded, the constitutional institutions of the country ignored the magnitude of the danger of anarchy or felt intimidated by the agents of terrorism; they did not take into account the impending threat to their own survival or the fact that if the ship of state were blown up, all the legal institutions and the autonomous foundations and associations of science and scholarship under the protection of the Constitution would collapse . . . Under such circumstances, the Turkish Armed Forces decided on behalf of the Turkish nation to assume the duty of guarding and protecting the Turkish Republic, assigned to it by the Law of the Armed Services, and carried out their duty within the military hierarchy. They took over the government in order to protect the integrity of the country and the nation, to safeguard the rights, the laws, and the freedom of the people, to secure their lives and property, to provide for their well-being and happiness, to re-establish and maintain the sovereignty of law and order, i.e. the impartial authority of the state, to reinforce the principles of Ataturk and prevent their erosion, and to lay down strong foundations for a floundering democracy.

It may seem paradoxical, but the conviction that democracy is the best form of government for the country is widespread among Turkish officers. The difference in their outlook is that they would like it to be the kind of democracy valued in their own world, which is totally different from that of the civilians, and deeply rooted in their training and education, their martial lifestyle, and their way of thinking. Their democratic model combines discipline, proper organization, disregard for self-interest in favour of the nation and the homeland, cooperation, unity, and constructiveness. And this is the crux of the matter: officers instructed with notions (not quite as flexible and tolerant as those of civilians) about what the state should be like, how it should be run, its aims and purposes, the kind of policies and politicians it requires, are inevitably disappointed when they discover that those who govern the country do not share their views on the kind of democratic process they envisage.

THE OFFICER'S IMAGE OF TURKEY

From his years in the Military Academy until the end of his career, the officer acquires an image of Turkey and its social conditions which is very different from that of the civilian.

For one thing, the officer is far more closely involved in the history of the country, especially in the fall of the Ottoman Empire and the War of Independence, which he has studied and discussed in much greater detail. Hence, he comes to believe that such periods of decline should never be allowed to recur.

Secondly, there is the general image acquired in the course of his military training: he is greatly influenced by the 'threat evaluations' (a phenomenon common to all armies) in respect of his country faced with definite or possible threats on all sides. The hypothetical but detailed grounding for such assessments does not reduce his conviction that the threats are real. And when he turns to internal affairs, the general picture is no less depressing: while such a picture is undoubtedly largely based on accurate and realistic observations, and on informative assessments, it is nevertheless coloured by pessimism.

> Turkey is surrounded by countries eager for our land and ready to attack us at the slightest opportunity. There are all sorts of speculations on the ways of doing away with part of our country . . . Besides, certain forces within the country, some with foreign connections, are waiting for the moment when they can carve up our land. The communists, for instance, or the Kurds . . . Then there are the religious movements which, apart from the struggle between various orders, are pressing for theocracy . . .

Thirdly, accustomed to a powerfully centralized system in his military career, the officer expects to find a 'powerful state' and is disturbed by instability in the state administration, and appointments made with an eye to political or ideological interests. Generally he tends to blame the political parties for the weakness in the state structure.

Finally, he deplores ignorance in society:

Who can claim that our people are not ignorant? Unfortunately they are. We all know about the rate of illiteracy . . . It is quite possible to deceive such people by exploiting their beliefs and feelings, and turn them away from the principles of Ataturk.

This kind of picture restricts the commander's loyalty to democracy:

The dangers facing Turkey are quite apparent. Equally apparent is the ignorance of the people. So, if a political leader attempts to turn the people away from the principles of Ataturk, he will have no choice but to confront us, no matter how many votes he gets from the ballot-box. He may fool the people but he can't fool us.

7
Does Their Training Prepare Officers for Intervention?

At the Military Schools and Academies future commanders are never told, 'You will intervene if necessary'. On the contrary, as indicated in the previous chapter, there is an insistence that the army should stay out of politics.

> Commanders should direct all their efforts to ensuring the success of operations in battle. They should avoid letting political considerations impinge on their military duties. One cannot perform military tasks with words, politics, or by listening to enemy propaganda.

Nevertheless, the students do begin in the fullness of time to sense that there is a role they may play in the country's political life. As the duties of the army are explained to them in their classes and in lectures on Ataturkism, the Law of the Armed Services and the History of the Revolution, they learn the significance of the 'protection' of Ataturkist principles and the 'defence' of the Republic against domestic and foreign threats.

There is in fact little that they need to be told on the subject. All they have to do is to look back on the last two decades and they will see that the army has intervened three times. What they learn at the Military Academies is limited to an explanation of why and when such interventions became admissible.

'At the time I was admitted to the Military Academy, I knew it was considered normal that the army should intervene if necessary. My father constantly used to say. "Thank God for the army. We can rely on it to save us." I grew up with this attitude. But I could not fully understand the reasoning on which interventions were based and when intervention was essential.'

'And have you found this out now?'

'Yes, of course, I know the answers now.'

'When do you consider an intervention justified?'

'When there is an attempt to do away with Ataturk's principles.'

'I don't quite get it. It's too abstract for me. Give me a more concrete answer.'

'If secularism disappears and the country is threatened by a shift towards Islamic Law, if communist or separatist movements arise (to put it bluntly, if an attempt is made to establish a Kurdish state) and the politicians are too busy with their own self-seeking squabbles to put an end to all that, then I intervene. What is more, it is my duty to do so.'

None of this conversation with a student in his last year at the Military Academy sounded in any way like a threat. He spoke like a man who was giving a matter-of-fact account of when and under what conditions he would fulfil a duty entrusted to him, and who sincerely believed in what he said.

Required reading on Ataturkism at the Military Academies draws a pretty clear picture of the duties of the Armed Forces:

> The overall task of the Turkish Armed Forces is to protect and guard the Turkish Republic against internal and external threats, and this task should be considered along with the Dynamic Ideal of the Turkish State, the Ataturk principles. Ataturk has described the qualities and principles that must always be borne in mind for the strong, harmonious and co-operative functioning of all the institutions within the state, such as Republicanism, Nationalism, Populism, Etatism, Secularism, and the Revolutionary spirit, and demanded that these principles and qualities be implemented and protected. All these principles and qualities are meant to ensure a strong structure for the Turkish state, which must first be built on sound foundations. The foundation will be strong if it relies on Turkish heroism, Turkish culture, and unites the Turkish nation. The Turkish Armed Forces constitute the unshakable foundation of the state.

Once they have been taught in detail the Ataturkist principles and their meaning, the student officers are told in unequivocal terms that they have been assigned the duty of guarding and protecting them. The state established by Ataturk has been entrusted to them, and the Turkish Armed Forces constitute the FOUNDATIONS OF THE STATE. They are told that this duty derives from the constitution, and is clearly spelled out in Article 35 of the Law of the Armed Services ('The task of the Armed Forces is to guard and protect the Turkish land and the Turkish Republic as designated by the constitution'). In other words, the mind of the young man taught to believe that this defence and protection is legitimate is cleared of all doubt.

In attaining the Dynamic Ideal, the principles to be observed and put into practice by the Turkish state are the same as those that will be guarded and protected by the Turkish Armed Forces . . .

Ataturkism regards the Turkish Armed Forces as the foundation of the state and has invested it with the duty of guarding and protecting the Turkish Republic against internal and external threats, in whatever circumstances . . .

The Turkish Armed Forces which serve as the invincible safe-guard for internal and external peace, and the general security of the Turkish State, have never been used for any other purpose than to deter or eliminate internal and external threats . . .

Their sense of national and military ethics, and the fact that they are ready to exercise power, and prepared to do so if necessary, in a realistic and appropriate manner, and with good timing, must be seen as a guarantee that the Turkish Armed Forces will guard and protect the Turkish Republic against internal and external threats . . .

Particular attention is drawn to the importance of the Turkish Armed Forces in quotations from Ataturk included in the textbook compiled for the General Staff:

Whenever the Turkish nation has taken a step towards progress, she has always seen the army of her heroic sons in the vanguard of movements for achieving the supreme national ideal . . .

When I speak of the army, I mean the enlightened sons of the Turkish nation which really owns this country . . .

The following is an excerpt (incorporating statements by Ataturk) from the same textbook on the question of the internal enemies against whom the Turkish state must be guarded and protected:

In order for the Turkish nation and her Armed Forces to be successful, they must know who their enemies are. There are two forces that prevent Turkey from attaining the Dynamic Ideal. One consists of external enemies with colonial intentions who would not like to see us making progress. But . . . an even more harmful and destructive group consists of traitors who are likely to rise from within us. Such enemies cannot be sensible and patriotic people who are aware of the realities, but those who are stupid and ignorant, or evil and unpatriotic, or blind. By ignorant we do not mean the uneducated; there are many who are educated but hugely ignorant as there are those who are illiterate but wise . . .

As noted earlier, attention is drawn in all these textbooks to the need for the army to avoid political involvement. But there is also an explanation of what is meant by political non-involvement.

In the fulfilment of their duties, the Turkish Armed Forces must take great care to avoid political involvement, which reduces in any army, whatever its potential, its capacity for united action. The Republic has always observed this principle meticulously. All the commanders, officers, and NCOs of the Turkish Armed Forces must be trained to be aware of the difference between intervention and actual participation in politics, and between being outside and above politics in the course of the fulfilment of their duties and responsibilities, in line with loyalty to Ataturkist principles. Within the Turkish Armed Forces, commanders who contemplate an attempt to use military power to serve their own political ends are rejected by the military forces and prompt corrective measures are taken. No political groupings or allegiances are tolerated in the Turkish Armed Forces. These are characteristics and values approved by the Turkish nation and expected of the Armed Forces.

The message is extremely clear. First, a young man destined to become an officer must never be trapped by outside influences into establishing ties with a political party, and a commander must never use the army for political purposes; anyone who does so will immediately be thrown out of the army. Secondly, it emphasizes that an intervention carried out when Ataturkist principles or the country itself are under threat from internal enemies is one thing, and involvement in politics or acting as a member of a political party is another; the latter will never be tolerated. In other words, the future commander learns that it is legitimate for the Turkish Armed Forces to 'intervene when necessary', but 'any attempt to retain their presence in the political field will not be tolerated'.

The second lieutenant before me had just graduated from the Army Academy. Bright as a button, self-assured, he spoke with confidence of the future. One of five children, he came from a moderately well-off family in Adana. He was extremely happy, partly because he had ceased to be a burden on his family and partly because he had successfully completed his studies.

'Looking back, what would you say was the first and most fundamental thing that the Academy gave you?'

(After a brief pause.) 'It gave me a personality. Instead of becoming an ordinary chap, I have become a man who has a duty in life, important responsibilities and a broader outlook. I feel that I am ahead of many of my peers. I have gained self-respect.'

'What is that duty?'

'First and foremost, to defend the country, both at home and abroad. Perhaps you may not be able to understand the feeling, but at this moment I feel the country is in my care and I am its true champion.'

'Why you and not me? What is the difference between us?'

'No, I don't mean you're not a champion. It's a very hard thing to explain. It is perhaps that I'm so concerned with the future of the country. I have the feeling I'm more concerned with it than you are. In plain words, I consider myself important.'

'Do you think, on leaving the Academy, that you'll have a say in the administration of the country?'

'It's not that so much, it's an awareness that Ataturk founded this country under very difficult conditions, and that he entrusted the country and the task of defending it to me.'

'In other words, do you see yourself with the right to intervene if there are certain developments that you don't agree with?'

'Not in the way you put it – with the right. It is more an awareness that it is my duty to save the country if it is in danger.'

'Then you regard intervention in politics as justified and do not believe Ataturk's dictum that the only sovereignty is the sovereignty of the people.'

'No, that's not how it is.'

'Well then, how is it? Tell me what you have the right to do.'

'I never forget what our commanders constantly drummed into us: "For goodness sake, lads, don't get involved in politics, don't sink into that bog." I therefore don't consider interfering politically, not for a single moment. But then, if my country is threatened with division, or is faced with a situation like that before 12 September, suppose it comes to face a danger from communism, and the constitutional institutions are unable to deal with the situation, it will naturally be my duty to save the country which Ataturk has entrusted to me.'

'Aren't you, in effect, inventing a duty for yourself?'

'No, you can't call it inventing a duty. I have been assigned that duty. What you refer to as inventing a duty is a process that arises naturally from developments.'

'You will now go into active service. Are you confident?'

'Yes, of course. My self-confidence has increased in the army. I believe that has been my greatest gain.'

This was not just an empty remark by the young lieutenant. There is, in fact, a sustained effort in the Military Academies to mould a self-confident man from the raw material supplied, purging him of any doubts or lack of courage. This point is emphasized in ceremonial speeches, in articles and commanders' lectures, and even in textbooks:

> Have faith in yourself because you are children of an intelligent and able nation. Even when appearing at its weakest, Turkey has proved through its army to be the strongest. Your guide

is the great Turkish nation and its history, your path the path of Ataturk. Never forget that your code of honour is to serve the country, the republic, and the armed forces with honesty, courage and honour.

The Turkish Armed Forces constitute the steely expression of Turkish unity, strength and ability, and of Turkish patriotism . . .

Whenever the Turkish nation has wished to go forward, it has always seen its army of heroic children as its leader . . .

In addition commanders, in inaugural speeches at the Military Academies or in periodicals published by the Academies, school their cadets on the following lines:

Cadet officers who will safeguard our future.

You should be proud of yourselves as members of the Turkish Armed Forces who are the only guarantors of our independence and our nation.

Friends, the belief that the best statesmen have come from the Military Academies reflects the hopes of our nation.

You are the chosen few who have qualities superior to those from other schools.

Let us briefly return to the conversation mentioned a little way back.

'I have witnessed how you young officers are looked after like rare flowers, and your commanders are constantly boosting your morale. Have you never developed a superiority complex as a result?'

(No hesitation.) 'At the beginning one does say to oneself, "My goodness, aren't I wonderful!" But it's a feeling that doesn't last. Oh yes, there are some among us, as there would be anywhere, who have an exaggerated opinion of themselves. But the system very quickly brings them down to earth. But it also boosts one's self-confidence.'

'Well, aren't you told at school what you must protect and guard, and when?'

'What we must protect is very clear. We must protect the

Ataturkist principles and guard against their violation. We have often argued the matter of protection among ourselves. It's not a palpable matter of concrete instances. We simply use the words "the Ataturkist principles are in danger" and "the country is under threat". Naturally, no one can determine the exact conditions these words describe, and that's how we always concluded our discussions. We would say that one's duty is according to the situation. This is how things stand now. But you're wrong if you think we spend all our time thinking about this subject and discussing it. It wouldn't even take up one per cent of our discussions. It just stays at the back of our minds as something to be remembered.'

'What do you think is meant by protecting the principles of Ataturk?'

'For the cadet in the Military Academy, it's quite clear that protecting the principles of Ataturk means protecting the homeland . . . The conviction is that our society is generally ignorant and can be deceived by politicians in pursuit of their own political or individual interests. The dangers facing the country from within and without are quite obvious. We see ourselves as protectors of Ataturk's heritage. When you look back on the recent past, don't you see a similar pattern? Once you read about the reasons behind the interventions of 1960, 1971 and 1980, the duties of the army become apparent, don't they?

A subsequent conversation with two of the top-ranking commanders in charge of education confirmed that their methods had achieved their purpose.

'A soldier's sense of identity derives from his feeling of superiority over a civilian, from his understanding of patriotism, unchauvinistic nationalism and the finer points of authentic Ataturkism, and from his willingness to die for his country . . . Our emphasis is on the achievements and aims of Ataturk. For instance, we try to discover what Ataturk attempted to achieve by secularism or etatism . . . It is the duty of every officer to protect the republic. The country is an entity. There can be no discrimination on the basis of language or religion. My duty is to protect them. To elevate what is Turkish is also our basic duty, and protecting it is a matter of honour. We are extremely sensitive on these points.'

'Educating the cadets in Ataturkism has gained weight since 1980. What do you intend to achieve by it?'

'Instead of a fragmentary approach to the subject which seemed to lead nowhere, we took up Ataturkism as an integral whole. In this way it has become obvious that Ataturkism is the real ideology.'

'There are important differences between civilian education and your way of educating the cadets.'

'Of course there are. The civilian standards of education, conception of the state, loyalty to country and to the principles of Ataturk are all very different from ours. In fact, even in foreign armies the principle of patriotism isn't as strong as it is with us. It used to be strong in the Prussian army, of course, but unfortunately that led to military dictatorship.'

'There seems to be a great deal of difference between the civilian's world and outlook and those of the officer. Isn't that undesirable?'

'The civilian's view of the country should be the same as the officer's. The civilian must be taught the same values. Friction will continue unless a different educational policy is adopted. My aim is to do away with differences not by resorting to arms but through dialogue and compromise . . . I encourage my officers to follow this path, to take the initiative and explain their values and outlook to the civilians and to persuade them that their way of thinking is more correct . . .'

'The officers regard themselves as the protectors of Turkey . . .'

'But you have to take note of the fact that they don't assume such responsibility in every field. They're primarily concerned with the protection of national integrity, not, for instance, with tax evasion or financial malpractice. However, they believe that if they're vigilant of everything that happens in the country, they can warn the civilian governments and propose solutions, and in this way they will be performing their duty of protecting the country. The principal concern of the officers is to be on the alert against any division in the country or any threat to territorial integrity. The rest they leave to the civilians. Officers are enlightened people; they wouldn't think of intervening on account of an economic crisis . . . If the politicians did their job, there wouldn't be any interference.'

'How would it be possible to eliminate the unease in civilian
–military relations?'

'Establishing a dialogue would be the first condition for
improving relations for the protection of the country. In fact,
the National Security Board was the best means of achieving
that . . .'

The following interview with a top-ranking general was also
highly illuminating:

'We would not think of intervening unless civilian governments were dividing the country. It should be sufficiently clear
by now that the army has no intention of getting involved in
politics. We firmly believe that any army that is politically
involved is bound to lose as a result of internal strife. That is
why we don't stay in power for long . . . Another point that
should be noted is that at each intervention the army has come
out of it unharmed.'

'Don't you have a very bad opinion of politicians? It's unusual
to have so little confidence in people who are meant to govern
the country . . .'

'What's bad is what they do, not what we think of them. We
don't have a problem, they do . . . You can see for yourself
that the army has always led the way to progress. Our sense of
national ethics enables us to make positive use of our resources.'

THE ROLE OF THE TURKISH ARMY IN HISTORY

No other army is as faithful to its traditions, or as keen to draw
lessons from history, as the Turkish Army.

Future commanders are well instructed in the history of their
army as one which is not confined to the republican period but
which dates back centuries to the Ottomans, and is colourful and
heroic, though at times unfortunate. Indeed, an historical view
of the Turkish Army makes it clear that military involvement
in politics did not begin in 1960, and that from the beginning
the army has functioned either as the ruling power or as an
inseparable part of it.

A brief survey of the history of the army summed up from
the point of view of a young cadet in the Military Academy
would reveal the following picture. From the foundation of

the Ottoman Empire in 1299 to its fall in 1918, the army always played a very active part. The early history of the Ottoman Empire is rooted in the exploits of the Janissary Army which was the principal source of strength that enabled quick conquests on three continents. Like everything else in the land, the Janissary Army belonged to the Sultan and performed indispensable services for him. The Janissaries fought battles, collected taxes, and governed the regions that were conquered. As the boundaries grew wider, the Empire became stronger and the Janissaries richer with the spoils of the conquests.

The Ottoman rulers always felt it necessary to keep the Janissaries contented. For instance, they were given bounty by the Sultan on his accession to the throne, and raised their voices when given less than expected, or none at all. Unrest among the Janissaries grew under the reign of less powerful rulers when there were fewer conquests and fewer spoils. For one reason or another, five Sultans and forty-three Viziers were removed from power or killed by the Janissaries in times of revolt: for example Sultan Osman II, who wanted to abolish the Janissary Hearth, and Selim III, who planned military reforms, were murdered in 1622 and 1808 respectively.

The active part played by the army and the navy in dethroning Sultan Abdulaziz was the clearest indication that the element of military interference in politics had not been eliminated despite the suppression of the Janissaries. However, the Janissary uprisings of the past had been motivated by financial rather than political interests. The first constitutional monarchy was also the first instance in which the army was involved in the overthrow of the Sultan for political reasons.

Unlike the Janissaries, who were reactionary, officers of the new-style army, in the course of military modernization, were influenced by liberal ideas from the West, and led the way in progressive movements and innovations. With a totally different outlook they renounced personal interests in favour of patriotism and the salvation of the Empire. But they also became much more involved in politics as the pace of modernization quickened.

Following Abdulhamit II's accession to the throne in 1876, the Young Turks and Ottoman liberals proclaimed the First Constitution with the support of the army. Military involvement in politics increased even more when Abdulhamit closed down the Assembly on the pretext of the Russian War and re-established absolute power.

In 1889, Young Turks and students from the Army Academy and the Schools of Medicine and Political Science formed the Committee for Union and Progress, which soon also included members from the officer class. The Committee had started as a movement against the repressive rule of Abdulhamit II, and succeeded in overthrowing him in 1909 with the substantial support of the army. However, the following years are full of examples showing the disastrous consequences of political involvement on the part of the army.

The period between 1909 and 1919, when Ataturk left for Samsun to start the movement for Liberation, was one of the most difficult in the country's history. The army was divided by conflicting loyalties in support of the government and the opposition, and internal strife led to defeat in the Balkan Wars. The history of this period, read in great detail in the Military Academies and never forgotten by any Turkish officer, is full of events which show how the army disintegrated as a result of political manipulation.

All of Ataturk's speeches which were intended to distance the army from politics bear the mark of that period. Although at the time his words were unheeded, Ataturk believed that the army was effectively progressive in giving political direction to the state, but that it had to confine itself to military matters after having done its duty, and leave politics to the Assembly. His point in changing from uniform to civilian clothes when he launched the War of Independence was to demonstrate that political and military matters had to be kept apart. He took great pains to persuade the people that the War of Independence was not a military uprising.

The following extract from a speech by Ataturk at the congress of the Union and Progress Party in Salonika in 1909 is the most vivid exposition of his views on the subject:

> As long as army officers remain members of the Union and Progress Party, we can have neither a strong party nor a strong army. A large proportion of the officers of the Third (Salonika) Army are also members of the Party. They cannot be described

as particularly powerful figures and people will be put off from joining the Party by the exclusively military support. Let us resolve the matter once and for all. Let the officers who want to stay in the Party resign from the army and let us also legislate to make it illegal for military men to take part in politics or join a party.

Ataturk believed that an army that did not enjoy the support of the people could not bring about the changes necessary to the country. The lasting reforms of a true revolution could only be achieved with the support of the people. He demonstrated his point when he set foot in Anatolia by wearing mufti and rousing the people to action for the War of Independence. The role of the army in the War of Independence is clear: it was the force that created the Turkish republic, but it could not have performed that task without the support of the people.

LESSONS LEARNED FROM TURKISH HISTORY

The Turkish officer of today in no way resembles the officer of the Ottoman Army. In fact, he does not like such a comparison. But he draws certain lessons from studies of the history of the period:

(i) The Turkish Army has always had its say and place in the government of the country and, at the very least, has influenced developments. Finally, through its role in the War of Independence, it established the Turkish Republic.
(ii) By becoming actively involved in politics the Turkish Army was divided and, as a consequence, suffered the defeat of the Balkan Wars. In Ataturk's words, it paid dearly for its involvement in politics.

These two instructive points lead the officer to form the following conclusion for today:

The Turkish Army must not become actively involved in politics but must guard and protect this Republic which was founded at the cost of so much bloodshed, and must intervene if these principles are endangered. Once developments have been settled it must return to barracks.

The future commander's approach develops therefore along the following lines:

(i) Ataturk salvaged this country from the pile of rubble which smothered it, and put it back on its feet, creating something out of nothing. To enable his creation to survive, he carried out certain revolutionary changes and set up certain principles. Turkey's survival depends on these principles being upheld.
(ii) Ataturk has entrusted these principles and this country to me. To defend these principles is to defend the country.
(iii) My country is surrounded by dangers from abroad and fraught with dangers at home.
(iv) My people are ill-educated and may be misled by ambitious politicians.
(v) The democracy that Ataturk established as a goal is my goal, but if the Ataturkist principles, namely Turkey's independence and future, are threatened, it is my duty to resist the threat. Let no-one think this country does not have its champion.

The 35,000-strong Turkish officer corps does in fact expect and demand that our population of 50 million should see its viewpoint and think along the same lines. And so far it has sought to achieve this without much effort, the reason being that it regards itself, rightly or wrongly, as more self-sacrificing, better-educated, more perspicacious, in short, superior.

This indicates the need to bridge the gap between the separate worlds of the military and the civilian, and as long as this situation is not rectified, and civilians leave dangerous voids in the running of the country, new 'interventions' can be expected. When and where does this parting of the ways begin, between the military man and his civilian peers?

8
Isolation From the Civilian World

Ever since I began to travel abroad and take an interest in foreign armies, I have been curious to know what others say about their own. In particular, I could never understand how some concepts that are hotly debated in Turkey are taken for granted elsewhere and are never discussed.

A Turkish expression that particularly aroused my interest was, 'our glorious army sprung from our nation, from our own flesh and blood'. In the beginning it sounded wonderful to me, but when I never heard its counterpart in other countries, I became curious. In the course of researching the present book, I raised the issue with people in Germany and France. They could not see the point of my question at all, and asked: 'What exactly do you mean?'

I thought I had made a mistake in the translation and rephrased the question. They thought about it and came back with something like this: 'What can be more normal or more natural? The Turkish or German army of course springs from its own nation. You don't think our army comes from the French nation and yours from Greeks or Italians.'

I was also struck by the adjectives used in all the speeches, official statements, declarations, and even newspaper reports that referred to the Turkish Army: HEROIC, GLORIOUS, PEERLESS, GREAT, OUTSTANDING, and so forth. I found that Western European countries also praised their armies, but they chose more modest adjectives like *powerful, well-organized, capable*, and, what is more important, used them less frequently. I asked the head of the Psychology Department of the West German General Staff the reason for this. He said:

> As soldiers, we know the strength and capabilities of our army, we know how well equipped and trained it is, and finally we know its courage and heroism. So do our public. So there is no reason or need to use such adjectives. Another reason is that it's not the done thing in Europe, particularly in Germany, to praise the army. Armies are needed when the chips are down, and we know that when they are the public will back us up. We have not the slightest doubt about that.

I posed the same question in Moscow. The Soviet public is proud of its army. They believe it is their greatest bulwark, and that the country's best guarantee for peace lies in the strength of its army. In addition, the great official propaganda machine constantly harps on the army's role in the October Revolution and the Second World War. Yet they also observe one important detail: they always present the army as part of the civilian population. They do not talk about 'the great struggle of the heroic Soviet army', but 'the great struggle of the heroic Soviet people'. Despite their trust and faith in the army, the Eastern bloc takes care not to put the army in the forefront. Although the importance of the armies in these countries is well-known, there is a reluctance to make frequent use of hyperbole.

I have not had the opportunity to study the situation in Asian and African countries, but in the Western world, the United States indulges in praise of its army more than Europeans, though less than the Turkish people do. The difference, however, is that the American public will, when necessary, criticize its army severely and in minute detail. There is probably no other country in the world where the army is the subject of such extensive debate in the press, in Congress, and among members of the public. Everything about it – from its failings to its choice of weapons, from its expenditure to its operations – is subject to the most minute public scrutiny. A typical outcome in recent years was that which followed the Grenada operation when press and TV correspondents were left behind on the grounds that there had not been sufficient advance notice. This resulted in the decision that, in any future operations, press and TV correspondents would accompany the first wave, irrespective of how secret that might be.

Armies do come under criticism and their affairs are subject to close scrutiny in Europe, but not to the same extent as in the United States. As for the Soviet Union and the Eastern Bloc, such an approach is conspicuous by its absence.

Having ascertained all this, I was curious to know how well the Turkish people knew the army 'it respects, cherishes and loves'. One of the reasons I began writing this book, in fact, was the feeling I had that the civilian Turkish public did not know the Turkish officer very well. The Turks are not so different from other people, and it is only when they find themselves in a tight spot that love for the Armed Forces takes on a new lease of life. The heart of the man in the street swells with pride as he watches his country's splendid soldiers on parade, sees his country's naval vessels sailing past and witnesses his country's air displays. There is a feeling of infinite confidence in this institution, the best organized in the country and one that has hitherto functioned without a hitch.

Yet, despite its close ties with the Turkish civilian public, the army is a machine that operates outside it. The difference between the civilian and military worlds is obvious. It is possible that Turkey uses such flattering adjectives in an effort to convey to its people the qualities and virtues of its army. The civilian population regards the military as a separate group. One reason for this may be the character of the army, but another may be the capacity of the civilian population to adapt more rapidly to change while, like all armies, the Turkish Army is prevented from adapting to social change by its innate conservatism.

When you look at various sections of the country, you can see differences in attitudes to the army. For the majority, as in the rural areas, and for the section of the population below a certain income level, the army is truly a great school of discipline, which has taught many a father and son to read and write and even follow a craft, filling their bellies with food of a quality and quantity they had never experienced before, raising their social standing and giving them security. Moreover, they respect the commander. The middle and upper sections of society take a more critical view, even though they are unable to express it openly. Unaware of how the army trains its officers, they are inclined to regard the soldier as a man who can only talk of heroism. Their unwillingness to criticize comes from their concern to 'avoid upsetting an institution that runs smoothly' and from 'the possibility of being misunderstood'.

In my view, the basic reason for the divergence lies in the civilians' ignorance of the most discussed institution in the country, an institution that influences the Turks' daily lives, i.e. the army.

The gap between the civilian and military worlds begins from the day the cadet sets foot in the Military School or Academy and continues until his retirement. The civilian population from which he originates and the retired officer who has to suffer the difficulties of re-entering the civilian world to which he is a stranger are both unhappy about this.

The officer who has developed a different logic and different value judgements has difficulty in establishing a dialogue with a civilian. The civilian, in turn, knows that the boy who used to play football with him is lost as a friend once he is an officer. They rarely keep in touch. When they meet again, they do not understand one another. The officer sees the civilian as a person who lacks the discipline and ideals he himself possesses. The civilian, in turn, sees the military man as someone outside the real world, with an inflexible commitment to a different set of values. Yet, when you think about it, they are moulded from the same clay.

So how do they end up as they are?

I shall never forget the first time I went home in uniform. That was the first time the 'difference' became clear to me. I was the same, but the boys with whom I had once played ball and engaged in friendly horseplay now saw me in a different light.

The first symbol of this man's apartness from the outside world was his general appearance. He had arrived home on weekend leave with his hair well-trimmed and his clothes well-pressed. Henceforth, he will always be similarly turned out on his home-leave until he has passed out of school. He will no longer be free to play tag in the street, join the local football team, or chase girls. His uniform must always be immaculate. His erstwhile mates, on the contrary, will be free to run around in blue jeans just as they please, to wear their hair long or short, or grow a beard or moustache as their fancy takes them. This is how the imperceptible parting of the ways begins. Their daily lives make it inevitable.

As soon as I entered the Academy I was told, 'you are different from the civilians'. That is something none of the cadets is consciously aware of in their first year as they are too busy trying to adjust to the new conditions but, towards the end of the year, the message begins to sink in. And the interest shown by the family on visits home – that is something else again! They praise you in words they have never uttered before simply because you have achieved an ambition on their behalf. The praise gradually transforms your position at home. On the Academy front there is the excitement of instantly making friends. After all, the people with whom you spend six out of every seven days in the week, and with whom you share your food, studies, and dormitory, have a greater impact on your life. So, by the end of your first year at the Military Academy, contacts with former friends dwindle and usually disappear over the years. From now on you are taken over by the new world you have entered and cut off from civilian life by the new environment and the uniform you wear. The pace of this process accelerates in the following years and you are absorbed into this other world.

While outside appearance creates the first 'difference', disparity in attitudes follows immediately afterwards. According to the rules of his school, the military student is being asked to do exactly the opposite to what is the normal practice among civilians; he must not talk in a loud voice, must show respect for his elders by getting up when they come in, be serious and dignified, and above all act in a disciplined manner, for example when carrying out an assignment or keeping an appointment at the specified time and place. The truth is that these are qualities that should be expected of everybody but in the context of Turkish society are regarded as being out of the ordinary.

After a while, I found it difficult to take the noisy arguments and invective-laden conversations of my civilian friends. I found it incomprehensible that grown-up people like them, university students to boot, could still push and shove each other while walking in public and how they could goad passers-by or, what is worse, harass women.

The next important difference between the military and civilian students stems from their different aims and motivations. It is perhaps true that the basic aim of both sets of youths is to succeed in their professional lives. The most important motivation in the civilian world is to make money, or at least to master some skill or profession that will ensure a comfortable life. You may describe this as a 'skill that speeds up the way to success'. To the youth who gains entry to university, personal success is extremely important. And the aim of the university is to turn that youth into a top doctor or top engineer and to equip him with the knowledge that will provide him with a comfortable living.

Are we not all told the same thing by our respective fathers? 'My boy, work and acquire a skill that will enable you to earn money. Otherwise, you will languish in poverty. Be aggressive, stand up for your rights, fight for them if necessary. You have to fight to get anything.'

The goal is to choose a profession with good earning potential and to succeed in it. In stark contrast, the young man who embarks on a military career is completely removed from these considerations. His earnings are pre-determined. He knows that, however successful his performance, he will not earn more than his classmates of the same rank. His satisfaction, in other words, will never come in financial terms. Knowing this, his instructors and commanders stress the other aspect of the military career: 'Money is not of the least importance. Yours is a profession that cannot be bought with money. On the contrary, you are men inspired by chivalry who strive not for money but for their country.'

The principles most frequently emphasized and repeated in the military schools appear in the civilian schools as unconvincing admonitions like:

You must be honest – Don't be self-seeking – Your country comes first – Cherish your colleagues, and defend their rights when necessary – Be respectful to your superiors – Make no demands, assignments are given only to those who are capable of carrying them out.

Finally, the greatest difference between civilian and military educations is evident in knowledge about the country. The cadet, as we have seen, is nurtured on Ataturkist principles, and taught that this country has been put in his care, and the idea that it is his duty to protect and guard the country takes root in his mind. He is educated for an eight-year period in subjects which range from the state of Turkey to responsibilities he may take over in the future, and is trained as the protector of his great heritage, the state.

The youth who attends civilian school, on the other hand, hears a far briefer version in classes of only two hours a week. More significantly, in Military Schools the teachers are army men themselves, brought up to take these principles very seriously, and overseen by an administration that also attaches great importance to their teaching. Given the time and attention devoted to teaching these subjects the military student absorbs them in their entirety, but the civilian student looks upon these subjects, which he has taken lightly from the beginning, as useful in gaining credits and not to be taken seriously. From this arise the most profound differences between the civilian and military students at the end of their studies.

Another difference in the Military Schools is the hard work and disciplined pace demanded of the students. Obedience is a cardinal rule. By contrast, the civilian student is left free to decide for himself and take the consequences.

When I met friends who were university students, they would suggest we went out somewhere in the evening, and attend to our studies next day. I couldn't go along with that, knowing that if I left today's studies till tomorrow, I'd never catch up, for tomorrow would bring its own work load. What's more, university students had no sense of responsibility or, if they did, it wasn't obvious to me.

A university student of the same age has a very different view:

My brother was at the Military Academy. He was used to such hard discipline that his life revolved around study and more study, and as the years went by, he began to look on me as completely irresponsible and undisciplined. I tried to explain that I might go to a discotheque tonight and catch up with my studies tomorrow by going without sleep. But he had become like a machine in his habits, and felt that if he had not finished his studies by a certain time in the day he would be penalized by his commander. By my lights, he wasn't living but merely obediently studying his lessons in a disciplinary strait–jacket. He became steadily more unyielding and inflexible.

The other brother did not share these sentiments:

Whenever I came across any of my old chums who had gone on to university, and that includes my brother, however much I brought up the subject of Turkey and Ataturk, I couldn't get a decent response. I found myself with people without ideals or curiosity, who didn't care what was happening to the country or where it was heading. Those who were politicized, on the other hand, were leftists, or supporters of the Justice Party [the largest right-wing party in the 1960s] or clericalists. Those who thought about the country or discussed it did so only from their own viewpoints. After a while, for fear of finding myself ridiculed, I avoided these topics altogether.

So we have, on the one hand, a youth who aims at getting into a good lucrative profession, determined to achieve that goal by hard work, but instead skimming through his studies to 'get them over quickly'. He tries to enjoy life as much as he can and either discusses the country in the light of his ideological position or doesn't give a damn about it. And, on the other hand, we have a youth who is totally committed to the military profession, determined, passionate, stubbornly trying to prove himself and get ahead, full of Ataturk and his country, and committed to what he sees as the supremely important task of defending the fatherland. There you have the inhabitants of two totally different worlds.

The difference in the quality of the education widens the gap. A comparison of Military Schools and Academies with civilian high schools and universities shows that the former have much higher general standards of knowledge and academic motivation. The difference is surprising even among graduates who have followed the same curriculum. The inadequate standards in secondary schooling give rise to difficulties and lower standards in higher education. University teachers point out the weakness of their students' grounding and complain of having to spend the first two years teaching them what they ought to have learned in high school. The superior education of the Military School graduates, on the other hand, raises the standards in the Military Academies.

Touring the ultra-modern campus at the new Naval Academy in 1984, President Kenan Evren remarked that if the quality of education in civilian schools and universities were not improved, there would be a huge difference between the future military and civilian cadres and it would not be easy to avoid interventions.

If the civilian–military dialogue has any importance for the future of Turkey, the obvious educational differences must be taken into greater consideration, and the standard of state education must be raised without further delay.

Why are the two educational systems so different?

The first reason has to do with the number of students. There is a total of 9,137 cadets in the five Military Schools and three Academies. The Ministry of National Education, on the other hand, is responsible for 12 million pupils in 52,000 primary, 4,900 middle, 1,181 high, and 1,083 vocational schools, with only 400,000 teachers in charge. Its budget is only 450 billion TL.

The education of a lieutenant just out of the Army Academy costs the state 6,700,000 TL, while that of a university graduate costs only 600,000 TL.

The number of students per class in the Military Schools and Academies is 15–25 as opposed to an average of 300 in the universities. Due to insufficient resources university students run short even of textbooks, while, as noted in Chapter 2, military students make full use of such facilities as language and chemistry laboratories, TV-video systems, free textbooks, 4-bed dormitories and extensive library services.

The greatest handicap for the universities is the high numbers of students, the lack of educational equipment and the problem of resources. The military educational system has brought down to a minimum the waste of resources which, for the universities, remains a major problem.

> In Military Schools we plan ahead and can tell beforehand what our needs will be and how and when to supply them. Co-ordination with other Military Schools makes it possible for us to be informed about the demands and supplies for each one of them.

The situation in civilian schools, on the other hand, doesn't bear talking about. But I could not help inquiring about it in a conversation with one of the top officials in the Ministry of Education, who replied:

> In our Ministry there is no long-term planning. Unlike the military, we don't know how many secondary schools or teachers we'll need in ten years' time. At present we're planning three years in advance, and although we are trying to develop five-year plans, I doubt if we'll succeed . . . In our system, the average number of pupils per teacher is 40. How productive can you expect that to be?

The difference between the two educational systems is not confined to the number of students. Other factors contributing to higher standards in the Military Schools may be summed up as follows: (i) preliminary elimination on the basis of academic achievement and motivation; (ii) assessment of personal qualities; (iii) assiduous monitoring and compulsory work; and (iv) the quality of the teachers.

Teachers in the military system have the same background in training and are committed to set standards. School administrators, preferably chosen from among staff officers, also undergo a special training and do not interfere with the teaching.

In state education, it is even difficult to find the right number of teachers, let alone teachers of quality. Teachers' standards vary according to their educational background, and despite the honourable work they do as civil servants, they suffer great hardships. Military teachers are given a good training, and are well provided for as regards accommodation and health insurance, while state teachers, with two children, who have served 25 years are on an average monthly salary of 135,000 TL, and must fend for themselves as regards housing once they are posted to the provinces. According to the statistics of the Ministry of Education, only 17 per cent of the 400,000 teachers are provided with accommodation.

In short, the differences in the quality of the students, of the teachers, and of the educational facilities account for the fundamental discrepancy between military and state education. They are also responsible for the divergence in outlook of the graduates and are an important contribution to the officer's sense of superiority over the civilian.

The civilian attitude also plays a part in distancing the officers from their own world. For instance, as civilians we would normally address an officer not by his name, but either by his rank or as 'Commander'. Some may be pleased by this, but many, evidently, feel uncomfortable about it. In general they do not consider it disrespectful if they are addressed by their names.

As a gesture of respect, or out of a desire not to offend, officers are not normally kept waiting in queues when shopping, civilian vehicles give way to military transport, and villagers in a coffeehouse stand up as a high-ranking officer passes by. No director general in the civil service would receive similar attention.

Furthermore, civilians are not inclined to air their views openly in the presence of officers, unless they know them well. And deference in an officer's own household increases as he rises in rank. Thus, the way in which officers are generally treated in Turkish society, by its very nature, brings about a distancing.

The difference is most conspicuous when comparing the circumstances of a university graduate with those of a young officer just out of the Military Academy. A graduate without support feels that he has no choice but to accept any available job, which usually turns out to be one that does not require much expertise or responsibility. The lieutenant, on the other hand, generally finds himself in charge of a squad of 30, or a company of 100–150. His responsibilities, which include his men and highly expensive weaponry and ammunition, have an immediate effect on his way of thinking and conduct. He is made aware that any mistakes he makes at this stage will influence his career.

'Why does the officer think he is different from the civilian?'

'He is brought up to a sense of discipline and responsibility. He is accustomed to working within a hierarchy, knowing what his duties are and how he should perform them. He is well-educated, well-organized and considers himself superior when he encounters the confusion in the civilian environment . . . For him, the Armed Forces represent supreme strength, honour and a single ideal. He feels distinguished.'

'Is he also influenced by his duty as protector of the homeland and the principles of Ataturk?'

'Of course he is. He feels that in this respect civilians are not as self-sacrificing, and he looks down on them.'

'Do all officers feel the same or do some think of themselves as equals?'

'You can't establish that from statistics. But generally speaking, some consider themselves far superior, others moderately superior, and still others equal to civilians . . . As with any section of society, it is difficult to be absolutely definite in such evaluations when dealing with the Armed Forces.'

'Then why does your educational system reinforce a sense of superiority in your officers?'

'It's a way of boosting morale. They must feel bound to an ideal. Their future will not turn out to be all that glorious, for they will have to serve under very difficult conditions. According to our internal regulations, for instance, we work on a 24-hour basis, during which our men do not complain if they are asked to carry a load of weapons or sleep on the ground. We have no choice but to boost their morale.'

'Isn't your educational programme far too intensive?'

'It has to be that way.'

'Why?'

'Because we don't have the financial means to train the same type of highly specialized officers as in the American or European armies. Our officers have to be all-rounders who can work in any field.'

'But wouldn't that mean officers without in-depth knowledge?'

'We do provide for specialization in some highly important areas, but on the whole we would like our officers to operate on a broad basis. It's a question of financial resources. We're not a rich country like the USA.'

'Aren't you troubled by the military–civilian divide?'

'Of course I am. There's no proper dialogue between us. Turkish society has a liking for its army but that's not enough. We each in our own shell watch the other from a distance. This is one of the major issues taken up by our Chief of General Staff. Now we're doing our best to open out, but the nature of our work does not permit us to be as forthcoming as you would like. Isn't it also an indication of our policy that we have provided you with the information you required for your book? But you must understand that an army cannot open out as much as the press or the universities. We're dealing with highly sensitive subjects which involve secrecy.'

'You said that it was important for the military and the civilians to come to an understanding. What steps have you taken to achieve that?'

'We shall try to make our views more widely known. In aiming for our country's progress, both sides are in fact trying to do the same thing. So it shouldn't be too difficult to reach a point of agreement.'

The words of the general in the conversation quoted above reflect the prevalent attitude of the General Staff.

Part Two
THE COMMANDER IN THE FIELD

9
At Your Command, Sir!

The first-lieutenant in front of me, had, earlier in his career, joined the most popular corps in Turkey, the Fifth in Thrace. In view of a possible attack from Bulgaria, aid from NATO has favoured this particular army corps and, with an eye on Greece, the Turkish government has endowed it with some of the army's most modern units, including the very special 54th Armoured Regiment.

On his way to Corlu to take up his posting the lieutenant reflected on its proximity to Istanbul. Many of his colleagues had drawn lots that had sent them to the Third Army's Ninth Corps, stationed in the remote Kars-Ardahan region.

'Having drawn Corlu as my posting was not at all an occasion for rejoicing. By the time we had left the Army Academy, we were so bursting with ambition and energy that it was almost a matter of honour to draw the toughest postings. We had no thoughts for our comfort and none of us coveted Istanbul or Ankara. Such considerations begin to take hold later on.'

'What was the most thrilling moment at your post?'

The first-lieutenant smiled the self-mocking smile of one who once held a high opinion of himself but has since seen the light.

'It was when I took over the squad entrusted to me . . . My unit of forty to fifty stood for inspection. Sergeant major, NCOs, the lot, all standing to attention. Right behind them, their vehicles. If you have been to Corlu at all you'll know the impressive field facilities of the 54th Regiment. Everything was quiet . . . And then I heard, for the first time, the wonderful words addressed to me: "AT YOUR COMMAND, SIR . . .". Suddenly I was on top of the world. I saw myself so great and

113

powerful that I thought mine was the ultimate achievement. It
didn't take me long to sober up, though . . .'

The first ten years or so of an officer's life (when he progresses
from second-lieutenant to captain) are the most arduous, and the
heaviest burden rests on the officer in charge of a detachment
of troops. It is his duty to train the recruits assigned to him,
accompany them under the most difficult conditions, be on
the go from five o'clock in the morning until lights out,
be on watch round the clock five or six times a month,
conduct night training once a week and take responsibility
for everyone under his command and for everything they
do.

Just a glance at the daily timetable is enough:

0500–0700:	Reveille, preparation, breakfast for the troops
0800–1230:	March to the training area. A five-km run, with rifle, in 17–24 minutes, and training according to the day's programme.
1230–1330:	Lunch
1330–1530:	Training
1530–1615:	Weapon maintenance
1615–1700:	Sports/Pentathlon
1700–1730:	Regimental evening inspection
1730–1900:	Supper and rest period
1900–2100:	Evening classes
2130–2200:	Lights out.

The lieutenant's work is not over when his men have gone to
bed. He is responsible also for other jobs, and he hits his bed
like a sack of potatoes, exhausted.

Many a lieutenant finds himself confused, panicky and disillusioned soon after arriving at his post. Throughout his school
years he was nurtured on the dream that 'if he bears up patiently
he will become a commander, the greatest and most sublime
office in the world'. He had lived in extremely comfortable
and almost luxurious circumstances, never having to concern
himself with his daily needs, for his every need was catered
for by others. A whistle dictated mealtime, another meant it
was bedtime. He had got into the habit of taking orders. And
now, as if propelled by a spring wound up over the years, he

had come to his command with all his ideals, ambition and competitive spirit.

In the first few days he had begun to say, 'I won't be able to do this job.'

There was no similarity between what I was taught at the Academy and what was expected of me at my post. If the commander had suddenly asked me what Star Wars was, or questioned me on its dynamic calculations, I could have happily discussed the matter for hours on end. But when I first came to my field command, I suddenly found myself in an environment and with problems with which I was totally unfamiliar. I had terrifying responsibilities. I had to be in charge of everything to do with the men under my command, from training to hygiene. It was my duty to take care of their food and train them how to use the toilet facilities. Suddenly I had acquired terrific authority. I was in charge of four tanks costing 550 million TL each, some 5 billion TL-worth of equipment, ammunition and vehicles; I had under my command men for whom I was responsible and, above me, a commander who expected instant execution of his orders. It was a most difficult period.

Of course the second-lieutenant has the sergeants and other NCOs to lighten his burden, but his lack of experience results in great tension. As the lowest-ranking officer in the whole army, the orders, queries and responsibilities that come down from the top always end up with the second-lieutenant. When a commander in the field issues an order, it is the second-lieutenant who has to carry it out in the rain.

These difficulties arise from the fact that at the Military Academies teaching is more theoretical and not much time is devoted to teaching the cadet how to administer men and units. This situation has gradually begun to change, but the problem will persist until the young officer is sufficiently prepared at the Academy and afterwards.

Second-lieutenants who serve this period under capable and understanding commanders, who enjoy teaching their juniors and prefer to correct rather than penalize errors caused by inexperience, sail through with ease. Those who draw a perverse commander, on the other hand, are either overcome by a complete loss of confidence that they will manage the job or are shocked into becoming warped or problem officers. The damage, in either case, lasts for years.

The first to test the second-lieutenant are the crafty old sergeants and NCOs of the team who are hardened by age and experience. They are skilful judges of a man's worth and grade their new commander according to type. It does not take them long to find out if he is shy and introverted, or hard and ambitious. They analyse his true personality in a matter of days and henceforth either 'manage' him as they see fit or accept him fully, placing themselves under his 'command'.

Having been evaluated first by his commander and then by his sergeants and NCOs, the second-lieutenant then undergoes the same process from his other ranks. The latter, of course, include not only illiterates but also secondary school graduates, educated, knowledgeable and sharp-witted city boys who approach him in the most innocent manner with the most innocent questions. He soon realizes that the ranks are summing him up, and want to find out how much he knows. Is he prone to fly off the handle? How does he react when faced with new situations?

> God forbid I should go through those first few months again. I felt like a fish out of water. Perhaps it was just bad luck, but I had the devil's own time. Maybe it was for the best, though, because I gained experience quickly. It was the pressure-cooker method.

One of the frustrations of a young commander is when the men misunderstand his instructions and fail to carry them out with the discipline that was the norm at the Military Academy. Still glowing with the memories of his time there, he wants the same standards of discipline, precision and hygiene. He is annoyed by the failure of the rank and file, who believe they will never have to wear a soldier's uniform again, once their 18-month stint as a conscript is over, to share with him the spirit of the Military Academy.

So one day you are faced with a fuming officer, ready to explode.

We left the Academy where a splendid discipline had prevailed throughout the ranks. We had been brought up as knights of idealism. I had rushed to my posting, excited at the prospect of creating the Turkey we dreamt of, and training disciplined soldiers who thought only of their motherland. I was ambitious but inexperienced. I realized that I hadn't been well or even adequately trained in the 'management of men', and inevitably I was disillusioned and frustrated.

It should not come as a surprise to hear these words from a general repeated verbatim by young second-lieutenants today because, although there have been substantial changes, the problem has not yet been fully solved.

During this initial period, the young second-lieutenant brings to his team the competitive atmosphere of the Military Academy. The effort to achieve first place without fail leads to an acceleration of the work pace, and also generates tension, stress and frustration among those not used to losing gracefully.

When he graduated from the Military Academy, his commanders had clearly defined his motivations for him:

You must apply endless energy to train the men around you to work as a unified team. You must strive to instil a spirit of nationalism, patriotism, and the consciousness of unity and solidarity in the men entrusted to your command. Ataturkist principles and fundamental goals constitute the basis for all your actions, and the basis of being a soldier is discipline.

*   *   *

All that was very well, the second-lieutenant thought, but how on earth was he going to apply this discipline with the men before him on parade? He was not at all happy with the way his commands had been executed of late. He had a feeling they were mocking him behind his back. Not that he hadn't made some mistakes, and he was aware of the subtle smile of amusement from the NCOs. He felt that the company was slipping away from him.

What made him finally blow his top was an incident the night before. On an unscheduled night-time inspection he had caught the soldier on sentry duty by the tanks napping, with his rifle propped up beside him. He felt the world collapsing around him. The scene ran counter to absolutely everything he had learned about the meaning of the Armed Forces, and he could not refrain from delivering a kick at the sleeping sentry. The latter, who had dozed off fatigued by the day's training, jerked back to consciousness and knew immediately what was in store for him.

The second-lieutenant immediately paraded his company. He thought to himself: 'They must be taught a lesson right away. If I don't do what needs to be done, they will make a fool of me.' He was of medium height and slightly built, but he gathered himself together and dealt the offending soldier a resounding blow. The sound of the blow had barely died away when there came a totally unexpected remark from the senior NCO:

'Sir, I would rather you hit me than him.'

The NCO was a giant, six foot four tall. A blow from him could easily flatten the lieutenant. But the matter had to be resolved there and then. What is more, the public statement was a challenge to his manhood.

'Alright', he said, 'come here and I'll show you a blow.' And he delivered it with all his might. The NCO paused, smiled slightly, and then said:

'Thank you, Sir. I deserved that. *Now* you are my Commander.'

*   *   *

The use of violence is strictly prohibited under the Law of the Armed Services and is defined as a statutory crime. If any victim of a beating complains, the commander of the platoon or company is immediately punished. The punishment, in turn, affects all promotions in that person's career. For instance, an officer given a prison sentence by a court is passed over for promotion for one year.

Army Regulations favour the lower ranks but for a number of reasons the rank and file soldier does not generally file a complaint. First of all, he is wary. Furthermore, research indicates that the soldier who is beaten up does not react as a soldier would in the same circumstances in a Western country. For one thing, 'beating' is institutionalized in Turkish society, particularly with children and even with youths; proverbs like 'spare the rod, spoil the child' are used to promote beating as a means of setting young boys on the path of righteousness; and, finally, being beaten is regarded as a routine occurrence.

Although beating is not prevalent in the Turkish Army, in the eyes of the men a beating is a form of punishment for sound reasons such as an extremely serious offence or an act disruptive of discipline. Studies also reveal that troops prefer punishment in the form of a blow or a moderate beating – odd, but true. A senior NCO explains (and I quote him verbatim):

> If I were to follow the rules, it would be necessary for me to hand over to the law any soldier who sleeps on sentry duty, fails to obey an order, or goes AWOL. That would mean confinement to quarters or a court sentence of imprisonment – a most undesirable outcome for the conscript, as his conscription is extended by one day for every day of his imprisonment, and results in delaying his demobilization. As far as the conscript is concerned, the loss of the weekend pass is worse than a good beating.

Why are they forced to hit or beat a soldier?

> According to the law, if a soldier has, say, sneaked off for a night, we have to initiate legal proceedings to have him tried by a court. This sets in motion a series of steps that takes months to complete and by the time a sentence is finally handed down, the offence in question has been forgotten. In other words, the sentence has no deterrent effect. By contrast, immediate punishment in the form of a beating is quite effective.
>
> What is more, if a platoon or company commander hands over one of his soldiers for trial, this creates the impression that he is not doing a good job himself, and that he is facing acts of indiscipline – not at all a good thing. So, while there are no hard and fast rules, in some cases a beating is the best way of dealing with minor offences, and resorting to legal proceedings for major ones, such as theft, desertion, causing a breach of the peace, ideological activities, etc.

Another senior sergeant, retired, put his finger on an interesting point:

A large proportion of our conscripts come to us poorly educated, cowherds straight from the fields. And here we have the task of teaching them discipline and how to use rifles. The poorer their education the harder it is to discipline them in the initial period. Yet, within two months at the most we have them thoroughly straightened out, and then we have to deal with the educated ones who resist a little longer. They too finally see the light, of course. There is no other way out. Nobody likes beatings but there is no other way to establish discipline. You know our people, as soon as you act a little softly, they walk all over you . . .

The situation which the young officer, fresh from the Military Academy and full of beans, faces on arrival at his posting has also a certain bearing on the question of beatings.

The second-lieutenant comes here without a thorough knowledge of how to deal with people and of what awaits him in the company; so, to establish discipline and conceal his lack of experience, he has to rely to some extent on beatings. Some become hard and inflexible as a reaction to not finding the kind of discipline that was instilled in them in the Military Academy, others because of their inexperience or inadequate grasp of the administrative aspect of command.

All these details should not give the impression that beatings are a daily occurrence in the Turkish Army. The truth is that just as beatings at school and at police stations, even torture at the latter, have not been stamped out despite every effort, certain practices do still occur in the army despite legislation to the contrary. The reason I discuss these matters here is that while they are now less frequent they have still to be eradicated, and eradication depends on the development of Turkish society. What I have done is to take an interest in a subject familiar to anyone who has done his military service, and researched the matter a little further. Now, what are the causes of these different attitudes in the ranks towards beatings?

It is noteworthy that those in closest contact with the rank and file are most involved with the beatings, the corporals heading the list, followed by sergeants, warrant-officers, and finally the second-lieutenant. In fact the same second-lieutenant ceases to figure in these incidents when he becomes a captain. Moreover, once he becomes a captain or superior officer with HQ duties and ceases to deal directly with privates, he reverses his attitudes and becomes the greatest opponent of beatings. Nobody believes, however, that they will stop until they are replaced by a more effective substitute.

Do you know whom the private soldier likes most? The strictest sergeant or company commander, the one who administered the most beatings. It's an odd thing, but I have seen with my own eyes that the soldier beaten most often cried his eyes out when the strictest sergeants were leaving the company. One thing the men look out for is that their superior (be he corporal or second-lieutenant) stands up for them. If they know they were in the wrong they do not hold it against the commander if he punishes them with a beating. And if the same commander also stands up for them although he beats them, then they worship that commander . . . In the rare case of a sergeant or officer showing a pathological tendency to beat his men, the news gets around very quickly. Once it is leaked to the base command, the person in question is finished. The General Staff are extremely sensitive on this subject.

An almost social incident . . .

I was doing my conscript service as a reserve officer. Soon after I took up my duties at my field unit, I began to get an electric shock every time I touched the telephone on my desk. I soon found out what was happening. One of the men, a smart alec city boy from Istanbul, had wired up the phone and, perched on a tree, used to laugh his head off, seeing me receive a shock every time I touched the phone. I sent word to the warrant officer to warn him off. My father was a general and had always told me: 'For heaven's sake, boy, don't beat anybody. Don't play around with other people's self-respect.' I had sworn to myself that I would not touch anyone. But the young chap's continued mischief was sending me crazy. Finally, one day, I called the unit out on parade and landed him a monumental slap in public. I was not proud of my action but it did the trick.

THE MOST DIFFICULT PROBLEM: TRAINING THE MEN

The rank and file conscript, known affectionately as Mehmetcik or 'Little Mehmet', is the backbone of Turkey and of the Turkish Armed Forces. Throughout the eighteen months of his compulsory military service he will perform any act of self-sacrifice without batting an eyelid.

Invariably, he regards the army as a school. On average, some 500,000 are conscripted every year. With the continuous fall in illiteracy, the special literacy schools within the Armed Forces were abolished in 1972, but even now an average of some 20,000 men are taught to read and write every year. There are even classes in Turkish for those who do not know the language. Some 92,000 are trained in 160 different skills (from driving vehicles to repairing machinery) that come in useful on their return to civilian life. When you add to these the 40,000 civilians employed by the army on contract, then the number of technicians and artisans trained by the army every year is in excess of 130,000. Bearing in mind that all the civilian technical secondary schools produce only about 100,000 graduates a year, we can appreciate the vast contribution of the Armed Forces in this respect.

Young men of all sorts leave their homes and come to the army from the four corners of the country. When they finish, they take back with them unforgettable memories as well as certain new ideas and value judgements ranging from Ataturkist principles to the meaning of the motherland and their loyalty to it. Some of these prove impossible to assimilate, others are soon forgotten, but the commander and barrack-room friendships are not easily forgotten.

One of the first lessons at the beginning of the four-month basic training period concerns the proper use of the toilet. Blocked toilets are a frequent occurrence, so the conscripts are taught not to use stones to wipe themselves clean:

> Look here, my son! You will aim for the hole and hit it dead centre. You must not take stones or anything like that with you;

you will find toilet paper and water there. That's what you should use for cleaning yourself.

The second lesson concerns bathing. It is an absolute rule that conscripts bathe at least once a week, and more frequently at garrisons with more extensive facilities. Then comes the proper way to eat, and, finally, how to use the dormitory, how to make one's bed the regulation way, washing and shaving in the morning, and so forth.

For some of the conscripts, these matters represent milestones in their education and stay with them for the rest of their days. They find in the army things that they had never actually seen or even heard of before. For many, this is an interlude in which they have a balanced diet of 3,500 calories a day and receive necessary medical or dental care. It is very much as if they are receiving their first lessons in a civilized way of life.

The proportion of conscripts who can read and write is much higher nowadays and the number of junior and even senior high-school graduates among them is on the increase. Nevertheless, for most of them the army continues to be a kind of school. This may be why a dim view is taken in Anatolia of those who have not yet done or are unwilling to do their compulsory military service. In some areas they will not even be considered fit to be prospective son-in-laws, so high is the regard in which the army is held.

Basic training is followed by assignment to specialist units, with the brighter ones retained as assistant instructors who will introduce new arrivals to the army's ways.

In the squad or company which is the smallest unit of the Turkish Army, the ranks are as follows:

(a) The officer (company commander), lieutenant or chief warrant officer.
(b) The sergeant (a better educated recruit, such as a graduate of a senior high school), the corporal (generally a bright but less well-educated recruit who undergoes a two week-training course).
(c) The privates.

An officer's greatest difficulty arises from the fact that he is expected to teach these privates everything from using a rifle to, eventually, manning a tank – all in the space of twelve to fourteen months after basic training. The existence of better educated individuals with a more highly developed learning capacity does help, but it is a tough task nevertheless. A particular difficulty lies in driving home the message that victory in a modern war depends more on the efficient use of high-technology weapons than on heroism and self-sacrifice.

Nevertheless, the conscripts' training is largely theoretical. For example, a soldier fires a mere 216 rounds in a year; each one costs 270 TL. Both privates and reserve officers know their rifles inside out and can take them apart and put them together again blindfold. But when it comes to hitting the target, they are less successful. Only two to six real tank shells – each costing 90,000 TL – may be fired in a year. And the reason is very simple: the country does not have the resources for more, and simulators to replace training with real shells have not yet gone into service.

On average the Turkish Armed Forces hold 11 separate exercises a year and 12 combined – with all three forces participating. In addition, they participate in 12 NATO exercises. As each costs some 10 billion TL, there is little prospect of increasing these numbers. There is, however, a daily training exercise in each squad. A relentless campaign is maintained to make the best use of available resources and to fill the gaps in field training. Nevertheless, the quality of training has yet to reach the level aimed at by the General Staff.

The private soldier is considered an extremely important person, particularly by the high-ranking officers. He is addressed as 'my son', and his welfare and protection from any unjust treatment is of serious concern. The following interesting observations on the relationship between Turkish officers and their men come from General Pendleton, for long the head of JUSMATT (Joint United States Military Aid To Turkey) in Ankara:

Turkish officers want to modernize their weapons system mainly for two reasons. First, of course, to have the effective means to repel any aggression. The second, however, is to spare the soldiers under their command. They are well aware that if they have to fight a war with the systems at their disposal today, their men will suffer heavy casualties because of their ineffectual weaponry . . . The thing that impressed me most during my stay in Turkey was the devotion of the commanders to their men. I have not seen the same in other countries. A Turkish commander will not sleep in a heated tent if his men are sleeping in the cold, probably as a matter of conscience . . . He takes care to share the same conditions as his men. It was most interesting to note that while the F-16 project was under discussion, the Turkish General Staff were pushing forward another issue: they wanted the technology for the manufacture of special cold-proof boots.

Teaching the conscripts what he himself learned at the Military Academy, the commander keeps watch over the constant process of change in his charges. At the end of their first two months, they have been licked into shape and are sent off to their units. It takes about six months for the conscript to become useful, but he reaches real proficiency and begins to operate at peak efficiency only in the last six-eight months before he is demobilized – and that is not long enough. As we have pointed out all along, the General Staff's current top priority is the training of junior officers and the men. This is because the weapons modernization will make the achievement of desirable goals even harder on the basis of the present intake of men, of their training, and of current practices. As there cannot be a full resolution of this problem without a substantial rise in the general level of education in Turkey, many foreigners – for which one should read NATO – suggest that the possibility of a professional army should be seriously considered.

Every Turkish male graduate from a university or its equivalent does compulsory military service for sixteen months as a reserve officer. The first four months of this period are spent in basic and specialist training. They then draw lots for posting and many are assigned to command squads. And this is where most of the trouble starts. Of course there is a difference between the performance of squads trained by fully-qualified officers with four years of Military Academy teaching behind them, and those trained by reserve officers for four months training only. The institution of reserve officer is a device to make up for the shortage of army officers, but many have misgivings about the degree of training given to them. Similar problems are known in other armies, too. One of the most important reasons for opting for a professional army is the fact that all officers at this level in the American and British armies are required to be regulars.

Reserve officers are not regarded as very reliable by the regular professional soldiers, which is not surprising, as their training and the worlds they belong to set the two groups miles apart. The professionals do not regard the reservists as sufficiently disciplined; they don't think they regard military service as the ultimate act of dedication to the motherland, transcending everything else, and they believe that they regard their army stint almost like a bout of 'flu', or an 'unavoidable accident', and that they count off the days till it ends. The professionals are substantially right in these assumptions: reserve officers who perform their duties with the same degree of enthusiasm and dedication as their professional counterparts are virtually non-existent.

As a result of this difference in outlook, reserve officers are not held in high esteem by the regulars, who avoid excessive fraternization with them. In the past, before the training of officers had reached its present level, this difference between regular and reserve officers was further complicated by a difference in their levels of education. The reservists used to regard the regulars as less well-educated, less knowledgeable or even uncouth, and this, in turn, used to create hidden tensions. While the situation has now changed, the fact remains that reserve officers are seldom invited to the homes of regulars or to discussions in the course of service in the field. A certain distance is always kept. It is all a bit like a club which excludes temporary members when an important meeting is being held.

MARRIAGE: THE FIRST ITEM ON A YOUNG COMMANDER'S AGENDA

When a young officer takes command after four years of the highly-disciplined life at the Military Academy he experiences a general sense of relief. First, he has proved himself by making the grade as an officer and, more importantly, the limited pocket money he received at school is suddenly replaced by an income of unprecedented size. Moreover, he can now dress in mufti. The requirement at the Military Academy that he should be in uniform even when off duty no longer applies.

He is now in heaven. First, he helps his family if they need help and then he begins to spend money on himself. Not that he has much occasion to do so. In this initial stage of his career, when work occupies him from dawn to dusk and any leisure activity is limited to visits to the Army Club, what immediately comes to mind is marriage. Marriage comes especially early to land force officers posted to Anatolia; they are spurred on by the concern that after the years at the Military Academy they may be plunged into loneliness all over again. The truth is that the unmarried are not made so welcome at the gatherings of married officers, or at least do not receive many invitations. So, if the young officer remains single, he finds himself sharing the table with other bachelors.

The young commander of twenty-two to twenty-five seldom considers the serious long-term nature of marriage – that the person he chooses as his wife will influence his career and play a role in his advancement, particularly once he has attained the rank of general. By the time he realizes that the superior qualities of an officer's wife – particularly educational attainments, command of one or more foreign languages, and a sociable but respectful nature – may favourably influence promotion it may be too late. The army – and even more the smaller community of the air force or navy – does not have a charitable attitude towards divorce, and assumes that there must be something wrong with the divorced officer or his ability to choose well.

The choice was even more haphazard in the past. Now, at least, a substantial proportion of officers are keen that their wives should have a job – but only while the husband is still in a lower rank and only a job that is compatible with the honour of an officer's wife. The wives are thus expected to make an increased contribution to the home. In fact, while marriage does initially appear as a solution to the problem of loneliness, it soon leads to new problems when the officer's pay turns out not to amount to very much. As a result, an increasing number of officers want their wives to work.

There is a gradual change in officers' wives which intensifies as their husbands rise in rank. Some wives (in common with the majority of Turkish wives) let themselves go, thinking, 'Right, then; my place is in the kitchen and the nursery', and there they stay. The truth is they are either unable or reluctant to see that with every passing year and every rise in rank their husband is forging ahead and increasingly opening up to the world. The upshot is that when the lieutenant rises to the rank of colonel and turns to look at his wife, he sees a woman who is still where she was when they first met.

But the interesting thing is that, as her status rises with that of her husband, the same woman is transformed into the most conservative stickler for the rules of rank. Many a colonel's wife would resent a major's wife, irrespective of their circumstances, walking through a door ahead of her, or speaking up before being given permission. She seems to want to preserve her social status, earned or not, at all costs.

Defending and exploiting the privileges of rank is much more common among officers' wives than among their husbands. The place of women in the world of the military is quite an extraordinary one. With some couples, the husband's failure to reach a higher rank causes greater anguish to his wife than to himself. It is perhaps a normal human instinct.

10
The Rise of a Commander

The lieutenant who finished the Military Academy full of ambitions and ideals undergoes yet another change when he is promoted to the rank of captain at the end of a trying ten-year period of field service. His years as lieutenant have toned him down and made him see that realities are quite different from what he was taught at the Academy. He has discovered what Ataturk meant by his statement that 'real training is acquired during field service'. He is more experienced and mature, and has stable and comfortable working relations. For instance, he has become more understanding towards reserve officers and less watchful of experienced NCOs. He no longer slaps conscripts, because he will have progressively less contact with them in the future.

With his promotion, the physical exhaustion of the past ten years is replaced by administrative responsibilities. These increase, especially during the average period of twelve years from major to lieutenant colonel when he is commander of a battalion or serves in the regimental headquarters.

Having left the toughest period of field service behind them, captains and majors discover how far they have moved from the world of books. Regimental service leaves little time for reading. Television news is more popular than the newspaper. Every regiment has a library, however modest, but the popular pastime in the barracks and officers' clubs is card-playing. After graduating from the Military Academy, an officer is given a chance every seven years to renew his interest in educational training in the Staff Academies, but the number eligible is limited. The fact is that after the initial training at the Military Academies, most officers cut themselves off from education.

Giving up books and reading, however, can have a negative effect on their professional training and make it difficult for them to adjust to world events and the conditions prevailing in the country.

The position of a 'commander', regardless of whether he is a lieutenant or a top-ranking general, is an institution in its own right. Even a squad commander enjoys privileges which other officers do not.

> I can never forget how much I enjoyed the years I served as company commander. I had 400–500 men under my command. I was even given a jeep and a driver, which seemed incredible then, and I slept in a separate room. Whatever might be the orders of the regimental commander, what really mattered for my men were my orders. It was I who decided how and to what extent his orders should be carried out. That's why no-one can ignore the lower-ranking commander. He holds supreme authority over his men. The moment you give orders you can set everything into motion: tanks, armoured vehicles, anything. It's a completely different world.

Whatever his age or rank, his additional authority and responsibilities involve a certain amount of risk in the way in which he carries them out.

> The majority of officers generally prefer field service to office work in headquarters. The way your men and the public look up to you indicates an immediate change of status. You become a different person. There's even a change in the way you walk. But it also involves stress. A telephone call from the commander of the division, battalion, or regiment is enough to give you a pain in the stomach. You start worrying if anything has gone wrong or if something has happened to one of your men.

The risk of taking the wrong step grows as responsibilities multiply. The commander is held answerable for all the decisions he makes and everything that bears his signature.

> I was company commander in Thrace, when the worst possible thing happened: one of my men disappeared with his rifle during

nightwatch. If a commander were to report this sort of thing to the police to have the fugitive tracked down, it would mean a terrible failure on his part. And if the soldier were to cross the border to the Greek side, that would mean downright scandal. It would signify that a commander had no hold over his men, which would be as good as a death-blow for him.

The status of a commander brings a wonderful feeling of satisfaction. And this means so much more in the provinces, where the people don't bother about the commander's name but regard him as the supreme regional authority, e.g. as the 'governor of Corlu'. He will have an adjutant, NCOs, and a chauffeur to serve him, and no-one will dispute his orders . . . While enjoying these privileges it becomes apparent whether or not he will make a successful commander. If his head is turned by power or status he will soon begin to draw criticism which will be filed in his record by vigilant superiors. Briefly, it's a highly satisfying but draining experience to be a commander.

Are officers trained for such posts? Do they attend courses in leadership?

For some 'commanders are born, not made'. The idea is that, however well he might be trained, the way he operates as a commander depends on his personal qualities. Others think that 'if an officer is trained in human psychology and management and if he has the right qualities, then the two factors will combine to make an excellent commander'. This is still a matter of debate in many armies throughout the world. In the Turkish Army there is some but not enough emphasis on courses in leadership and psychology. Leadership still depends on the officer's natural disposition. However, commanders are always under close scrutiny. Their domestic life, physical fitness, conduct under stress, whether they are irritable or cool-headed, use bad language or not, are points on which they are judged by fellow officers and privates. Such value judgements form an unwritten code.

The qualifications and status of commanding officers in the Turkish Army have undergone a great change over the years:

In the old days, even in the 1960s, a commander was like a god: hard, dead serious, unsociable, terse, but with a ringing roar

. . . In time, however, as educational standards rose among the graduates of the Military Academies, the reserve officers and even the privates, this type of attitude on the part of a commander was regarded as a smoke-screen to conceal his ignorance. I'm not saying that commanders were ignorant then, but simply that they were not as knowledgeable as they are at present. Nowadays expectations are much higher: first of all they must be well qualified not only in military expertise but also have higher standards of general knowledge and culture. The popular commander now is one who takes a special interest in the day-to-day problems of his men, not who gets his way by yelling and shouting. The good commander can tell what his men want from the look in their eyes. Those who fail are marked as 'ignorant', and it's difficult to shake off this stigma. A commander may enjoy much satisfaction, but he is also a target for criticism.

A reading of Army Regulations shows that there are now guidelines on how a commander should treat his men. Instructions generally tend to favour subordinates, in order to curb arbitrary behaviour on the part of the commander and to encourage him to be protective towards those under his command. The penalties for ignoring such instructions are not as light as one might think. However, in practice things may be quite different. As always, the human element, that is, the personal qualities of a commander, determine how successful he is.

INSPECTION: THE FEARFUL DREAM

To a commander, inspections represent the same kind of anxieties as examinations to a student. The difference is that examinations do not take place so frequently.

Once, when Semih Sancar was Chief of the General Staff, we got news that he would be coming down for an inspection of the transport regiment in Malatya. The first thing was to find out what he paid particular attention to during his inspections. After some enquiries we discovered that he would first inspect the kitchen, the food, and the provisions . . . So, immediately an extensive clean-up operation was started in the kitchen, where everything was painted white and all the pots and pans polished. We were very pleased with ourselves when Sancar did indeed go straight down to the kitchen, as predicted. However, at dinner that evening he smiled and said: 'It was the same situation in my time. We used to clean up before an inspection but once that was over, things would go back to what they were before.' We were terribly upset that he had seen through the whitewash.

Inspection is one of the most important events recurring in a commander's professional life. As it will inevitably affect his future, he tries to follow the moves of his superior from the moment he leaves his headquarters. A well-placed friend at headquarters is the ideal source of information so that the proper strategy can be devised.

Every commander knows from personal experience how the system works. So some favour surprise inspections, while others have made it a principle to spring on impromptu inspections in addition to those at fixed dates. The latter are kept under close surveillance because no-one in the army likes to be taken by surprise.

Special care is taken to note the commanders of regiments, brigades and divisions, and corps commanders, whose movements are the most unpredictable. The army commanders or Chiefs of the General Staff do not pose a problem because their movements can be closely followed.

Some commanders are known for their special interest in morning exercise or general hygiene, so conditions in a particular unit will reflect that interest in anticipation of an inspection. But a large proportion of inspections are of a different kind which involves registering details which generally escape notice.

During my inspections I used to enquire from the company or squad commander about the privates' difficulties: whether they got their mail regularly, or had health problems, etc. If the answers were inadequate, I would conclude that the commander did not take sufficient interest in his men . . . For instance, it's not enough that the artillery is kept brightly polished, what really matters is whether proper maintenance is carried out on the machinery. An experienced commander can immediately tell what is fake from real, although the unit under inspection never gives up trying to put on a good show. Some commanders are openly critical, while others may be more reticent. But no experienced commander is ever fooled. For instance, a glance at the privates as they go about their business is enough to convince the commander at a briefing session of the state of affairs in that unit.

The real test takes place during manoeuvres, where it becomes obvious how well each commander has trained his men.

RECORDS, PROMOTIONS AND APPOINTEMTNS

The strongest indication of an officer's professional progress is promotion. Officers who are promoted before their time are considered special. The military system of promotions is different from that of the civil service in that it does not require an officer to complete a fixed number of years before his promotion is due. This is the most important factor encouraging competition among officers to be the 'best' in inspection and to be the commander of the 'best trained unit'. Competition intensifies as officers rise in the military hierarchy.

The race for promotion up to the rank of colonel is quite different from that to the rank of general. To be promoted, it is sufficient for an 'unambitious' officer to complete the given period of service without any grave errors, violations of discipline, or shirking of his responsibilities. For those with higher aspirations, however, it becomes all-important to achieve precedence over others as quickly as possible. One way of achieving this is to train one's unit well; another is to ensure a good record. There is no hope of a bright future for an officer with a bad record.

The yearly reports for an officer's record are based on the average marks filed by three superiors. Full marks are 200 for officers and 100 for NCOs. If there is a discrepancy of 30 points or more between two sets of marks, both superiors have to submit written reports. Similarly, written reports must support the marks on any question regarding morality or discipline. It is the final average of marks for the total period of service for a particular rank that affects promotion.

The report forms consist of 20 questions for officers (10 for NCOs), covering personal, domestic, moral, and intellectual as well as military and professional matters. Until recently there also used to be a query as to whether the officer was 'patriotic', but this was deleted because patriotism must be taken for granted as the precondition of being an officer, and is also an abstraction that cannot be graded.

An officer's personal relations with his commanders can also influence the reports they file on him. Any resentment between them or difference of opinion on personal or professional attitude is bound to play a part. For instance, the query on whether an officer likes drinking may elicit a black mark from a teetotaller who has caught his junior with a glass in his hand, or a good mark from another who considers this a virtue in itself.

As elsewhere in the world, assessment by marks has always been a controversial subject in the Turkish Armed Forces. For want of a better system, measures have been taken to improve the questionnaire.

The significance of the recorded marks is that they have an important bearing on promotions, entry to the Staff Academies, and rises in salary. A captain or major must have 120 points, i.e. 60 per cent of the annual total, to get his promotion. In the case of lieutenant colonels and colonels this goes up to 70 per cent. To gain seniority in the same rank and be promoted one year in advance, officers must score 95 per cent in addition to having other assets.

The competition in some cases is so close that even a single point can make a difference. To improve things, examinations on 'general culture and basic military knowledge' have been introduced since 1985 for all ranks except lieutenant colonels and colonels. While one reason for this is to base promotions on safer criteria, another is to encourage officers to do more reading. But only time can tell how successful this will be.

The General Staff do not seem to mind the complaints they receive on reports:

We are aware of some of the hitches in the system. But they can't really make a big difference, because an officer may have several different superiors reporting on him during the year. Within six years of service in a particular rank, there could be 12–18 commanding officers filing reports on him. These shifts should prevent any instances of gross unfairness or favouritism. The few cases there may have been remain negligible.

Clearly complaints cannot be avoided unless the human element is eliminated and replaced by computer assessment. The problem, which exists even in the American and German Armed Forces, is as old as the armies themselves. Nevertheless, the general opinion of Turkish officers is that, despite all its inadequacies, the system of assessment is not unfair in practice. It is no great problem for an officer in normal circumstances to rise to the rank of colonel. Difficulties arise for those who want to get ahead of others.

No full picture can be given of an officer's life without reference to appointments, which play such an important part. 'Every appointment is like a house fire', is the saying among the troops. Indeed, there must be very few armies which keep their officers more constantly on the move.

'I've been in the service for 44 years now and I've had to move house 30 times, and send my children to 30 different schools, some of which were bad. I feel like a migrant', said a full general, complaining of the frequent change of appointments that Turkish officers have to cope with. Officers stay on average 2 years in one post, though in some cases this may vary from 18 months to 3 years. The frequency of appointments inevitably affects the officer's efficiency. It normally takes a year to adjust to the new conditions while the second year is dominated by worries about the next appointment. Appointments at the General Staff and headquarters of the three services are the most difficult, as it takes time to acquire the necessary expertise for such positions. As all appointments proceed according to regulations, an extension of the appointed period is not regarded favourably.

The river Euphrates has been taken as the boundary to divide Turkey into the first and second (West and East) regions. Appointments in the East mean serving in desolate regions which have no health care and poor educational facilities and where roads are closed for 7–8 months of the year. Usually, every officer serves twice in the East: first between the ranks of lieutenant and captain, and a second time between the ranks of major and colonel.

The Armed Forces are the only Turkish institution which has consistently maintained an uncorrupted appointments system. In other organizations, such as the police force, some ministries, and the state-run Economic Enterprises, an undesirable appointment prompts immediate action on the part of the person concerned to activate a network of friends and personal connections to avoid the change. In the military system similar attempts would only harm an officer's career. There may be some appointments influenced by personal preference, but in general the principle of equality is strictly observed in rotating appointments between the eastern and western regions of the country.

One advantage of the frequent changes in appointment is that they provide officers with the chance to become acquainted with every part of the country. Secondly, they prevent 'outsize' cadres or maintaining the same cadre of officers in the same positions or garrisons for too long. This would deprive the lower ranks of the chance of moving up and would lead to dissatisfaction; it would also allow for over-attachment to a fixed position.

Moreover, even the most influential Chief of General Staff or Service Chief is usually retired as soon as his time is up. This is probably one of the most important features distinguishing the Turkish Army from those in Latin America or other developing countries. Under exceptional circumstances the hierarchical pyramid may become overcrowded at the top because it is possible, professionally or politically, to extend some generals' periods of service. But these extensions are never for long. The purpose is to allow for some flexibility in the rules so as not to force into unnecessary retirement a general who is worthy of his position.

Following the 1960 political intervention, overcrowding at the top resulted in the summary retirement of 5,000 officers. It would be difficult to find another army capable of such action in order to preserve its fitness. In the course of this study, no other army has been found which could reduce overcrowding, or eliminate officers not committed to the required principles, with such speed. In our opinion, this is yet another factor that prevents decay in the Turkish Army.*

* In 1962–3, 1,800 cadets were expelled from the Military Academy for involvement in the Aydemir incident, and about 800 officers and NCOs were dismissed after the 12 March 1971 intervention. We have no figures for those dismissed before or after the 12 September 1980 intervention.

11
The Commander's Fringe Benefits

Members of the Turkish Armed Forces are reasonably well treated by the state in return for their heavy responsibilities, their dedication, the exceptional selflessness expected of them and the exhausting pace of their work. The present-day Turkish Army cannot be described as 'rich', nor can it be said that its officers 'are not adequately rewarded for their efforts'. The difficult periods prior to 1960, which gave rise to such remarks as 'would you want your daughter to marry an officer?', are long behind us. A detailed comparison with staff in the civil service may justify the conclusion that the military receive better salaries than a good many sections of the bureaucracy. Such a comparison is, of course, extremely difficult to make and highly subjective. It involves such things as functions, skills, authority and responsibilities of the individuals concerned and the nature of their duties. We have therefore been careful to include in our scale of comparisons only those whose duties have reached a certain level.

The worst enemy of all officers' salaries is inflation. The levels to which their living standards had fallen prior to 1960 are well-known. The adjustment made following the 1960 revolution kept the situation steady for a time, but the sharp rise in inflation towards the end of the 1970s once again upset the balance to the detriment of the officers. As a result, a new arrangement was instituted, which, with minor adjustments, has lasted until now. After the 1980 military intervention there was a proposal for a new adjustment in military pay, but it was shelved following the objection by the military leadership that it would 'give rise

139

to the false and objectionable impression that the intervention had been carried out for the purpose of amending the officers' pay scales'.

Overall, the take-home pay of the Turkish officer is slightly better than that of the comparable civil servant, though the difference narrows with seniority. The real difference, however, derives from the fringe benefits that the army provides for its members.

HOUSING

Living quarters head the list. According to official figures, construction work is under way to provide living quarters for up to 70 per cent of the personnel of the Turkish Armed Forces. This will amount to approximately 50,000 units. Some 72 per cent of the target figure has already been achieved at the time of writing. Priority is highest in underdeveloped areas where housing is hard to come by and lowest in the major cities.

This construction, providing units of a minimum of 85 square metres and an average of 100 square metres overall was pursued until recently without proper standardization and programming. While one may occasionally come across sprawling units of some 400 square metres for the top brass in any of the three services, one could also come across officers of comparable rank in Ankara who have to put up with living quarters of 100 square metres. In 1986, however, the General Staff ruled that living quarters would henceforth be built according to standard sizes (85–150 square metres) and that the practice of each service carrying out its own housing construction was to be replaced by a central planning and programming board, controlled by the General Staff.

The rent paid by officers of all ranks for their army living quarters is 100 TL per square metre per month automatically deducted from the salary. Bearing in mind that civilian rents can be as much as six to eight times as much, particularly in the major cities, it is easy to see what an important contribution army accommodation makes to an officer's budget. Moreover, the availability of such accommodation frees him from the onerous task of finding a house to rent with each new posting every two years.

These living quarters are impressive in their simplicity and their well-kept condition. The military guard at the gate, the upkeep of the gardens and the generally high level of maintenance and cleanliness, all carried out by military manpower, ensure that officers are able to lead simple, comfortable and civilized lives.

Not everyone, of course, is able to secure a place in these quarters. Rank tells, as always. There are four kinds of quarters:

(a) 'Special allocation quarters' of 120–150 square metres allocated to lieutenant-generals, full generals, and their air force and naval equivalents. There is no waiting list for these higher ranks.

(b) 'Post allocation quarters' for major- and brigadier-generals and their air force and naval equivalents, as well as officers of a lower rank on critical or important duties. These are somewhat smaller units but, once again, there is no waiting list.

(c) 'Service allocation quarters' allocated to the civilian drivers of the General Staff, the Ministry of Defence, and the Service Commands; also to the resident concierges and household technicians of army quarters, and to personnel assigned to duties in border posts and communications centres.

(d) 'Miscellaneous allocation quarters' allocated to officers and warrant officers according to a complicated points system.* Not all can be provided with quarters, of course, and the budgets of those who are not suffer in consequence. The most disadvantaged are the junior and warrant officers.

---

* Each officer or warrant officer is credited with one point for each year of service, plus an extra point for every month from the day he takes up his duties, applies for quarters, but cannot be provided with accommodation. If his wife is not gainfully employed, he is credited with six points; if she is a state employee, with two points. Plus further three points for each child, one point for every other family dependant he has to maintain and two points if his family has no income at all, three points for every award for good service and, finally, fifteen points for service in the field. On the other hand, three points are deducted for every sentence received from a court or for suspension of promotion by a disciplinary tribunal. If he, his wife, or any of his dependants has a house within the province where he is posted, five points are deducted. This total number of points determines his place in the queue for the allocation of quarters.

There is no evidence of such a widespread practice in the West. In some countries (the United States and Britain, for instance) living quarters are allocated to officers in certain circumstances. Those who do not want army living quarters receive a rent allowance. The great difference is, of course, that the rent payable for army living quarters in the West is only marginally below the market price. Not only does Western practice avoid insulating the officer from civilians in his daily life, but it also contributes to the economy by providing him either with an allowance towards his rent or with credit to buy his own home.

This housing system is not exclusive to the Turkish Armed Forces. But, because the army is so much in the public eye in Turkey, their living quarters stand out, not least for their distinctive appearance, their brightly-lit entrances with sentries on guard, and also because they are so obviously well-maintained. All this gives the impression that the Armed Forces is the only establishment that builds housing for its members, whereas in fact, in the past ten years in particular, all sections of the bureaucracy (even members of parliament) have been busy building living quarters for their members. Moreover, many other institutions in Turkey, prevented by inept legislation from paying adequate salaries to their employees, find themselves compelled to build housing in order to retain their personnel. Yet there is no serious debate in the country as to whether this approach is saddling the state with an increasingly large and sterile investment burden. There is not even an awareness that there may be a problem here.

HOSPITALS

In respect of hospitals, members of the Turkish Armed Forces enjoy far more benefits than civil servants and even employees in the private sector. All forms of medical treatment in military hospitals – of which there are 42 – are free of charge for officers and NCOs (in service or retired) and their families, and for reserve officers and privates. Civilian patients are admitted only up to 5 per cent of their capacity.

In the case of house calls, only 20 per cent of prescription charges are paid by patients. If the required treatment is unavailable in Turkey, they are sent abroad with special permission from the Academic Council of the Gulhane Military School of Medicine, and the cost is met by the Armed Forces.

Conscripts are among the groups benefitting most from this system. Apart from the medical examinations at the beginning and end of their service, they are checked once a month during the first six months, and subsequently once every three months; those who are ill are sent to hospital. Health care is particularly important for those conscripts who come from underdeveloped areas; this may be the only chance in their lives to have medical or dental treatment.

Besides helping the family budget, military hospitals are commendable in the way they provide the necessary care and treatment without subjecting patients to the misery or indignity of state hospitals, even for patients in state health insurance schemes or for those patients who can afford to pay. The standard of medical staff in the military hospitals is as high if not higher than that in university hospitals, and great care is taken to maintain their standards.

In the event of death, the funeral expenses of active or retired officers, NCOs, any member of their families, or of conscripts, are met by the Armed Forces. Colonels and generals are honoured with a ceremonial funeral, unique to the military institution.

CLUBS AND STAFF RESIDENCES

Officers' clubs hold an important place in the daily life of military staff, with their special atmosphere dominated by a certain manner of speech and strict observance of hierarchy, even in respect of retired officers. They are yet another factor increasing the officer's isolation in his own world.

At present there are clubs in forty of the Turkish provinces, and entertainment halls in other provincial towns or small headquarters. The clubs offer an accommodation capacity of 2,500 rooms and 9,600 beds for officers and their families, in addition to other facilities.

In the official wording, officers' clubs 'were established to promote solidarity and cultural links among the members of the Turkish Armed Forces, to provide accommodation, dining, leisure and entertainment facilities for active and retired servicemen and their families . . .' In fact, however, their primary function is to provide the above-mentioned facilities at prices in line with the officers' family budgets, i.e. five to ten times cheaper than in hotels of comparable standard. The cheapest rate for a room is 150 TL, the most expensive (in Ankara) 950 TL; an average meal costs 500–600 TL.

In the Gazi Officers' Club House in Ankara, which is a much better constructed building than the Buyuk Ankara Hotel, most of the rooms on the first seven floors are wood-panelled and attractively furnished with minibars, TV, and telephone connections. On the eighth floor there are seven luxury though rather flashy suites, reserved for top-ranking foreign guests of the Armed Forces or the Ministry of Defence. There are also magnificent halls, covering an area of 178,000 square metres. Other club houses, two in Istanbul at Kalender on the Bosphorus and Harbiye, and one in Malatya, also provide accommodation for foreign guests, but are not as luxurious as that in Ankara. It would be wrong, however, to assume that officers indulge in luxury in their club houses. The average room for all except generals and admirals is of reasonable dimensions. In some provincial club houses a room may accommodate four people.

Observing hierarchy is equally important in the club houses. The dining space, for instance, is divided into three areas: for junior and senior officers, and for generals/admirals. Advancement in rank is immediately obvious from the way officers enjoy their change of social status in the club houses. It is considered unacceptable for an officer to socialize with his commander, unless personally invited to do so.

The higher ranks are given priority in accommodation. Rooms are available only for non-residents by permission, and it is generally not so easy for junior officers to find a place. Much depends on the resourcefulness of the commanding officer in charge of the club house. This position usually allows him to form useful connections, especially with higher-ranking officers.

In view of the low cost of everything from wedding and engagement parties to valet services, and barber/hairdressing facilities, the club houses are a paradise compared with civilian establishments offering similar services.

How was it that a day at a civilian hotel of the same standard cost 30,000–50,000 TL and a comparable meal at a civilian establishment came to 8,000–10,000 TL per person, yet it could be so cheap at an army club? My guess was that these impressive bargains were subsidized from the Turkish Armed Forces budget. The commanding officers of army clubs I interviewed said exactly the opposite.

> First and foremost, we do not pay rent for our premises. They are built from budget funds. Therefore we do not have capital expenses or pay interest. Also, we have no personnel expenditures. We use conscript manpower and the civilians we employ are paid by the Turkish Armed Forces. Minor repairs and, since last year, utilities such as electricity, water and so forth have to be paid for out of our trading capital. Where we score is in our finely-judged purchasing policies and our ability to pare costs to the bone. Then again, we avoid the high profit margins of commercial establishments. That and our relatively low overheads and side expenses enable us to keep costs down . . . The law requires that we should not operate in the red but avoid a profit margin of over 10 per cent . . .

To take one example, the Gazi army club ended 1985 with revenues of 260 million TL and expenditure of 217 million TL, that is to say some 43 million TL in the black.

For an officer, arrival or departure at the club is an experience to savour. The prompt salute he receives from the ramrod-straight sentries at the club makes him glow with pride in his rank and profession. Would he get anything like the same welcome at a civilian restaurant, even if we disregard the fact that he could not afford to dine there in the first place?

In 1970, when Memduh Tagmac was Chief of the General Staff, no civilians, including officers' families, were allowed into the clubs. They may now be admitted but only if they are members of an officer's family or guests of a general or admiral, and the presence of too many is not regarded favourably. Not every officer is able to invite his friends, presumably because of limitations of space.

Officers' clubs are also out of bounds to NCOs, who have their own much more modest clubs. This leads to grumbles: 'we are the backbone of the Army but we're treated as second-class members'. Aware of this touchiness, the General Staff recently initiated a number of measures, including pay adjustments, designed to upgrade the NCO clubs to the level of officers' clubs. The results can be seen in the construction of buildings on the Maslak highway in Istanbul, at Sihhiye in Ankara, and in Malatya and Erzurum. According to a high-ranking commander, who sharply rejects the arguments that the NCOs feel they are treated as second-class citizens: 'An NCO joins the army knowing exactly what he can and cannot do, and there is no question of any kind of second-class treatment'.

Army clubs are a hive of activity at night, with heated discussions, particularly among retired officers. The army club is a lifesaver for the retired officer. It provides him with the pleasure of cutting himself off, for a few hours at least, from a civilian world he has never been able to come to terms with and of being with comrades-at-arms who think and talk as he does. But the regular clientele consists mostly of those who have retired before reaching the exalted ranks of lieutenant-generals and generals. The latter pay few visits, no doubt remembering the days of past glory when they were waited on hand and foot and anxious to avoid being treated as 'old-age pensioners' now.

REST CAMPS

Should you, in your travels through Turkey, come across any beautifully well-kept but somewhat nondescript buildings in some lovely spot, with a sign that says 'SPECIAL TRAINING CAMP', you must not be misled into thinking that this is a place for weapons-training or anything similar. What you have stumbled on is a holiday camp for members of the Turkish Armed Forces. There are twenty-five so-called 'special training camps' in Turkey, with a total capacity of 12,266 beds.

The basic aim of these camps is to provide active and retired officers and NCOs with the opportunity of a rest. The move to set them up began largely during the period when military pay was low. Since then they have been the subject of much public discussion.

The move to standardize the camps has been a recent event. In the past, each service and sometimes certain individual units within them were able to set up their own camps. Thus, for instance, there is a naval camp in Antalya to which only naval officers go; other officers are not barred but nobody wants to go against the established custom. In the same way, the air force and the gendarmerie have their own camps in Cevizli and Canakkale respectively. These are camps largely geared to summer holidays. The only winter holiday camp is in Uludag. One camp is exclusively for NCOs while the others are mixed, admitting NCOs, but only up to 30 per cent of the camp's capacity.

The attraction of the camps derives from the combination of good service and low prices. Thus, for instance, a fortnight's holiday for a husband and wife can cost as little as 20,000 TL compared with 150,000 TL at the very least in a comparable outside establishment. But the entitlement to stay at the camps is only two weeks in the summer and one week in the winter, and is awarded on the basis of a points system. So the actual chance of spending a holiday in one of the camps comes once every four years on average, and as the choice is generally for June and August, one may have to take pot luck with dates. The generals may have a somewhat better chance but the junior ranks would be lucky to have a total of two stays in one of the camps by the time they have reached the rank of colonel.

OYAK AND THE PX SHOPS

There is nothing quite like OYAK, acronym for the Army Mutual Aid Association (Ordu Yardimlasma Kurumu), either in Turkey or anywhere else in the world. It is an institution to assist the regular officers and NCOs of the three services and the gendarmerie, as well as the civilian personnel working for the Armed Forces. Unfortunately, it comes in for criticism from some of its own members and is eyed with suspicion by civilians. Launched with a capital of 50 million TL in 1961, OYAK is worth 36 billion TL today, a gigantic institution with 88,000 members, sole or part owner of 28 industrial or commercial enterprises, and with a cash revenue of nearly 23 billion TL in 1985. It is a judicial entity, bound by special judicial relations and is financially and administratively independent.

The concept of OYAK was created when a way was being sought after the 1960 revolution to prevent the recurrence of the drastic fall in income of the members of the Armed Forces in the 1950s.

> We wanted to establish a system to ensure that members of the Armed Forces would not have to depend on the government alone for their income. We wanted to provide them with additional support while in service (for loans, PX facilities, housing loans), with benefits for their families in emergencies such as death or disability, and to ensure any member against penury on retirement as a result of an inadequate income from the Pension Fund. So, without any funds from the state, we developed a system based on deductions made from the member's payroll and putting these funds to work to provide social services. Later on, civilian organizations sprang up which tried to emulate OYAK, but they did not survive. Supplementary insurance organizations do exist in Western countries, but they are not quite like ours.

OYAK deducts 10 per cent from the basic salaries (excluding allowances and so forth) of its members. This money is put to work to provide services for the members, the most important being the lump sum paid on retirement. This draws the sharpest criticism from the members, perhaps because they are unable to understand the complete picture.

> We have 10 per cent deducted from our salaries every month and that is not something to be sneezed at. Then, when I retire, the sum deducted is returned to me with the accrued interest, calculated at the official rate of 5 per cent per annum. The fact is that interest paid by the major banks has been hovering around 40 to 50 per cent. I could get more money if I deposited this money in a bank. OYAK is getting cheap money from me and uses it to earn high rates of interest, and then returns it to me

with the interest calculated at the low rate. How can one put up with this?

The administrators of OYAK refute the allegations. They say they provide excellent dividends to their members.

We pay interest on the deductions at the legal rate of 5 per cent. However, we add to this a 9 per cent auxiliary interest and at the end of each year the dividend payments from our profits. In 1984, for instance, we paid a dividend of 28 per cent to bring the total net interest on our members' capital to 43 per cent. Banks were paying a net interest of 47 per cent during the same period. So we did not do too badly with our rates.

Members retort that dividends are not the same every year.

That's true, they are not. But that's because economic developments and inflation are not the same every year. In 1975, for instance, the dividend we paid to members was 2.5 per cent. It rose to 4.5 in 1980, to 10 in 1982, to 23 in 1983, 28 in 1984, and 24 per cent in 1985. Thus we ended up paying a 38 per cent per annum interest to our members in 1985 when the commercial rate of interest was 43 per cent. That's to say we paid a little less than the commercial rate. Those of our members who complain completely overlook the fact that their membership entitles them to extremely attractive opportunities, grants and loans. For instance, one of the most important of these advantages is the housing loan.

In the mid-1980s OYAK issued an average of 2,000 housing loans a year. Loans are provided to those who want to buy a house on the civilian market, and OYAK also builds houses through its own construction company and sells them to its members at a price considerably lower than that of the commercial market. A 100 square metre house costs 11 to 12 million TL, and some 2,500 houses have been built to date. Even so, it cannot be said that OYAK has satisfied its members to any great extent.

OYAK alters the maximum amount for each housing loan according to the rate of inflation. The latest ceiling is 2.5 million TL. The chances of buying one of the houses built by OYAK is nil, as demand is so high and the queue so long. As for outside prices, they are frightful. The maximum loan of 2.5 million TL is miniscule. Nobody will accept it as a down payment. You can get nothing except a jerrybuilt block in a very down-at-heel quarter. It is out of the question for me to live in such a place while I wear this uniform. What's more, I have to pay off the loan over a period of ten years, at 12 per cent interest per annum. That is not very attractive, particularly for junior officers. I've been paying 10 per cent of my salary for the past twenty years. One can buy a house only when one reaches the rank of colonel – that's to say, long after you have paid a sum equal to the cost of a house, and then only by paying all over again.

OYAK does not share this view at all.

If they want to obtain a housing loan from other sources, they may have to pay up to 35 per cent actual interest. The rate of interest charged by the Emlak Bank ['Property Bank', which acts as a kind of building society] and the State Housing Fund is 16 per cent for a flat up to 75 square metres, 22 per cent for 100 square metres and 35 per cent for one over 100 square metres. It is perhaps true that the loans we advance are too small, but then that is all our resources allow us.

To obtain a housing loan, one must have made contributions for a minimum of ten years, and priority in allocation takes into account seniority and the number of children.

To advance a loan to an officer, its own member, OYAK requires a guarantor. It is in fact a lot easier to make purchases on a ten-month hire purchase agreement outside. That is the extent of the OYAK's confidence in its own members.

Nevertheless, the OYAK member enjoys a number of advantages.

Grants in the event of death or disability in active service. So far a total of some 606 million TL have been paid in 5,278 cases.
Loans. Members may borrow a sum up to 1.5 times their salary, for a period up to two years, at 12 per cent interest. Some 130,000 persons have had loans totalling 13 billion TL so far. Repayments are regularly made from the payroll.

Then there are the PXs, the army stores. There are a total of 20 army PXs in fifteen provinces and districts in Turkey, including Girne (Kyrenia in Cyprus). The PXs were taken very seriously by the National Unity Committee (the junta that carried out the 1960 revolution) on whose decision OYAK was established. They first came into operation in 1963. Their prices are 15 per cent below those in the outside market. In addition, they sell durable goods like refrigerators, TV sets and washing machines on one-year instalment plans at 12 per cent interest (compared with 30 per cent outside). The popularity of the army PXs can be better understood if one takes into account that their sales were over 10 billion TL in 1984, with profits of over 500 million TL. OYAK's answer to those who claim that the PX shops could make much greater profits is that 'profit-making is not our aim; selling lower-priced goods with guaranteed quality is'.

There are many objections to this from within the army: that the shops are down-market, and that it is difficult to find good quality goods in them. Another point of contention was voiced by a retired full admiral who once was commander of the navy:

After forty years of service and in return for all the contributions I had made, I was paid off with a few million TL. While we were in service, there were occasions when we received 100 per cent pay rises – yet OYAK's loans have failed to keep pace with these changes. All right, let bygones be bygones. But how about the gigantic investments OYAK made with our contributions? In other words, there must be accumulated funds whose value is increasing year by year. I was, however, paid off and left to my own devices; but how about the investments of so many billions created from contributions from the likes of me – what will happen to them? Who will get them in the end? Then again, what should I do with the lump sum compensation I've been paid? It would be better if they had kept it and continued to use it to provide for our families.

According to businessmen, OYAK could be managed much more profitably. But the managers of OYAK say:

> We operate like an insurance company. Every insured person pays a premium which the insurance company utilizes and, when the time comes, pays out to the insured. No insured person, once he has been paid, is entitled to what remains of the investments. We provide our members with opportunities on terms they cannot get anywhere else.

One of the questions most frequently asked is whether OYAK is being operated as an institution which serves as an auxiliary source of income for the Armed Forces or as a source of jobs for the boys after retirement. The head office employs about 250 people. Our research revealed that seven of them were graduates from the Military Academies and fifteen NCO school graduates. Of the 500 personnel at the PXs, twenty had a military background. It proved impossible to find out the same sort of information about the 20,000 employees of the companies owned wholly or partly by OYAK. It is well-known that there are appointments of retired military personnel to the boards of certain institutions but OYAK officials insisted that these were all civilian institutions and that there had been no suggestion by the Armed Forces that there should be such appointments.

> High ranking commanders do not even want to attend our meetings or put in an appearance here, let alone make suggestions to us; they are averse to giving the impression that they may have something to gain. As for the officers who criticize us, they are taking out on us their frustration at not being able to criticize anybody in the service. Criticizing us gets rid of their pent-up anger. When we explain, they do see things our way. Our main problem is with the retired personnel. Having nothing left to do, they let fly at us . . . The day we let this place begin to look like a place for good pickings by the military, we automatically go bankrupt. The General Staff is also very sensitive on this point.

The only segment of the military who do not get their fair share from OYAK are the reserve officers. Throughout their service, 5 per cent of their salary (half the rate for regular officers) goes to OYAK. When they are discharged, the money remains with OYAK. When I looked into this odd situation, I was told:

We too are aware of this. And that's why we have cut down their contributions. What we are doing is insuring them against accidents and death during their military service. And we have increased payouts for these.

12
Staff Officers: The Cream of the Army

If you know Istanbul at all well, you are bound to know the Maslak urban highway which leads from Levent to Istinye or Tarabya. Every time I drove that way, I would notice on the right, in the hazy distance, a group of ultra-modern sombre buildings that did not appear to have any traffic around them. But I found that if you tried to park your car anywhere near these buildings you would immediately hear soldiers blowing whistles and sentries telling you that 'It is prohibited to stop'. You would eventually find out from surrounding notices (No Unauthorized Entry/Prohibited Area) that this was no ordinary place. Among the trees in this area of seemingly inaccessible and slightly eerie buildings a huge notice proclaims: 'THE STAFF ACADEMIES'.

I had some notion of the nature of the Staff Academies but, like most people, I had no idea of what went on inside them. I always had a vague feeling that they were the kind of site for top secret activities, a place that gives you a slight shiver as you go past.

The day I entered the Staff Academies, however, I came face to face with an entirely different world. I found myself in a campus where you can take care of all your needs without setting foot outside, a town of some 3,000 inhabitants where fresh-faced young officers with books under their arms dart from one building to another. I use the word 'town' advisedly. The Staff Academies occupy an area of 1,180,000 square metres, and constitute a self-sufficient town so well organized that it can comfortably cater for all the needs of its residents. There is provision for the needs of the working wives of student-officers: their children can be looked after from morning until

154

evening, fed and cared for, for an annual fee of 10,000 TL. Its other amenities include the most modern sports facilities, which practically compel one to use one's leisure for sport. You may be forgiven while you tour the Staff Academies if you forget you are in Turkey. I asked myself repeatedly, 'Were these all made here?' Yes, they were and, what is more, by the Turkish people. They form a town that arouses both admiration and amazement in the beholder.

There is no single institution, but separate Air, Land and Naval Staff Academies, because of the differences in their functions. The same mini-town also houses an 'Armed Forces Academy' and a 'National Security Academy'.

For a moment, the whole set-up looked to me like a factory which takes in a regular officer through one door and transforms him into a staff officer by the time it puts him out through another, a gigantic piece of machinery churning out some 125–130 staff officers for the Armed Forces every year.

WHAT SORT OF PERSON IS A STAFF OFFICER?

A senior artillery captain's answer to this question is quite interesting. I felt that he might have been describing an ideal man.

> A staff officer is a realist with ideals, who is forever searching but adapts to quickly changing situations; he has the spirit of enterprise, is always within sight of the commander but manages to stay in the background; a mature and modest person who understands the reasons for obstacles, a thoughtful man of few words who knows what he is saying and admits what he does not know; a person who has foresight and whose presence is less seen than felt.

That was the description of a person rarely found in Turkey. Another officer's definition of a 'staff officer' in relation to a commander is as follows:

> A staff officer is someone who is constantly inquiring and treats the results of his inquiries warily. A commander uses speed, a

staff officer comprehends speed. The staff officer is a man who seeks initiatives. The commander is the man in the public eye, the staff officer is the man in the shadows. The commander overcomes disabilities, the staff officer knows the reason for them. He is a man of few words who knows what to say but will not say what he doesn't know . . .

Well, why did they decide to become staff officers?

Staff officers have played an important role in Turkish political and military life, which has influenced the country's fate. I am here to serve more effectively in the army. Everybody wants to climb to the peak of his profession, and in the army that means becoming a staff officer.

The most important goal motivating an officer in the six years from lieutenant to captain is admission to one of the Staff Academies and becoming a Staff Officer. Staff officers make up the cream of the Turkish Armed Forces, and being one is the dream of every officer who wants to get somewhere in the army. They know that the road to the top – to the rank of general – means achieving staff status and that staff officers constitute the backbone of the army; for instance, 75 per cent of the 24 colonels promoted to general in the land forces every year are staff officers and the chances of making further headway in that rank are doubtful if they are not staff officers. Achieving staff status is so important that it may even cause family rows, with a wife yelling, 'You have still not been made staff officer; look what your classmates have achieved!' The public also perceives the staff officer as somebody special, a person with a future lying open before him.

In addition, an officer who completes the Staff Academy course is automatically credited with an extra three years of seniority, thereby leaping ahead of the non-staff officers of his graduating class. He gets a higher salary and progresses fast. His chances of a foreign posting also rise, as in choosing military attachés preference is generally given to staff officers.

The officer who is admitted to the Staff Academies is aware that many important doors will automatically open before him.

What he does not realize is what a tough course he is letting himself in for.

The fundamental task of the Staff Academies is to teach management and administrative skills to post-graduate level or, in other words, to fill a gap in the job done by the Military Academies by providing the trainees with managerial and analytical tools and teaching them 'situation appraisal'.

Here is how high-ranking commanders from the Staff Academies explain their goals:

> What we're doing here is training officers to know what's what and how to sort things out. A staff officer is a problem solver, a fixer. We are teaching a system that can put pieces of data together, scrutinize them carefully and warily and take an objective view without getting lost in detail . . . What we're supplying is the art of systematic thinking. We're teaching what we call 'situation appraisal'. The Military Academies turn civilians into soldiers and sometimes enable the latter to acquire some superior qualities. But the work of the Staff Academies is to teach the art of war.

They are not intended to turn out specialists. Trainees must know everything – operations, personnel matters, logistics, etc. About 60 per cent of the teaching during the two-year staff course deals with military subjects including the principles of war, tactics, strategy, the nature of various weapons, and the effects of the latest developments in military strength. The officer is also subject to management training and courses which raise his general cultural level.

> Over the past five years we have increased the number of classes on Ataturkist principles and on Turkey's general problems. We want our officers to know and understand their country well, and to this end we review Turkish economic, social and political developments and encourage the officers to discuss them, and bring in outside lecturers.
>
> The young man we train here will go to the General Staff or to one of the various army corps to take up important duties at headquarters. He will be regarded as a 'distinguished-knowledgeable-superior officer'. He will go on missions for

international organizations or to foreign postings. We are therefore trying to train him well beyond the present level of our army's weapons-systems. But we don't ignore the situation if he returns to his unit disillusioned.

The Staff Academies keep future staff officers well informed on information technology and the technologies of the twenty-first century as well. But doesn't this beg the question: aren't we trying to train a tractor driver as if he was going to sit behind the wheel of a Rolls Royce?

It's a valid point. Our officers here learn about Star Wars down to the minutest detail. They get to know US-developed air and land military strategy backwards. They know all about Germany's World War Two strategy and its outcome, but they leave all this information behind them. There's no opportunity to use this knowledge when they go to their units. And that's a fact of life in our country. Turkey hasn't caught up with its own age, yet we are training our officers for the future. That is what we must do. And when we've done that, the knowledge we've imparted remains theoretical and on paper. Unable to use it in his daily life, the officer begins to forget it after a while, unable or unwilling to keep on updating his knowledge. But even half of what we have taught him is enough to maintain him as an officer with excellent qualifications.

In the Staff Academies we found the same kind of heavy workload as in the Military Academies. The principle that the officer should know a little about everything, like a general practitioner in medicine, forces him to work hard.

The officer who enters the Staff Academy is pushed to the limit in the first year. He is up against a system he is not familiar with. Some may even lose hope that they will make the grade. In the second year, having come to grips with the system, he feels relieved and settles down. He has found out what he must learn and to what extent.
Another reason why we push so hard is that our trainee has difficulty digesting the training at the rate we have planned for him. He comes to us unprepared, having shut his books on

leaving the Military Academy, and when he became involved with field work he forgot what he had learned. The reason we want to teach him everything is that we are short of officers. Things are bad enough now but would be worse if we made them specialists. Being a poor country, we are trying to get on by teaching our men to perform a whole range of functions.

For many years there has been a shortage of officers in the Turkish Armed Forces. This has blocked the way to specialization and affected staff officers as well. Each year 120–130 staff officers graduate from the Staff Academies, which have so far produced a total of 4,800. Of these about 3,000 are on active service, but the numbers are still too low. The recent programme for the 'Selection of Reserve Officers' has been started for the purpose of recruiting graduates from technical universities to supply the need for specialized officers. (Under the same programme, the rest of the university graduates serve as conscripts for nine months instead of sixteen.) Specialization will be unavoidable in future due to the pre-eminence of new defence systems which require advanced technological knowledge.

FACILITIES

The facilities provided by the Staff Academies are impressive. The most important of these is the excellent library, unrivalled in any Turkish university. It is kept well-stocked with the latest international publications for the benefit of officers doing research. Courses can be conducted on closed-circuit TV. The computer centre, laboratories, lecture halls and classrooms are up to modern Western standards. The staff-student ratio (3:5) is in itself an indication of the quality of the studies.

The Staff Academies also provide accommodation for married officers in two- or three-room houses on the campus. Single officers and teaching staff are accommodated in the 130-room guest-house. The aim is to create an ideal environment for study. Nevertheless, future staff officers do not have an easy time during their two years at the Academy. Six hours of classes, followed by a lecture or seminar and evening study every day do not allow them much free time to spend with their families. An important factor affecting their work is the support they get from their wives.

The Staff Academies also admit officers from abroad. In 1985–6, for instance, two American, one South Korean, one Bengali, and four Pakistani officers attended the Academies. Foreign officers are admitted by agreement between governments and are provided with fully furnished houses, depending on the terms of the agreement.

As the main problem is language, foreign officers have to attend special courses in Turkish. Some courses, especially those considered 'secret' in view of the sensitive topics they deal with, are not available to foreigners. This seems to be the common practice also in the Academies of Britain and the USA. A qualification from the Turkish Staff Academies is regarded as a high distinction in South Korea or Pakistan. The presence of American officers in the Academies is mainly because of a special interest in Turkey. Following their period in the Academies they are generally posted either to JUSMATT or to American bases in Turkey. A knowledge of the language and, more importantly, friendly relations with fellow Turkish staff officers from the Staff Academies are considered to be an advantage.

The Academies are also responsible for effecting some changes in the attitudes of staff officers. Compared with the average officer who tends to think of himself as a specialist after a two-week course on a subject, staff officers become more respectful of careful and in-depth study, and more cautious in their analyses. They view events from a different standpoint.

Furthermore, they develop a stronger interest in political developments in their country. They are 'sensitized' to problems. The subject of Ataturkist ideology is treated in a more scientific manner in the Staff Academies. They become more involved in the realities of the country in the course of lectures and discussions on probable scenarios and plans of action.

There are those who think, mistakenly, that we train our men for intervention. Our scenarios are intended to cover all the possibilities which our officers are expected to consider and debate before arriving at decisions on particular situations.

In general, it is very difficult for officers to be admitted to the Staff Academies,* but once they are in, they rarely fail to pass unless they make grave errors. According to official figures, the rate of success is 99–100 per cent. The entry requirements are so strict that officers who are seriously involved in their work find it relatively easy to graduate.

During the two-year period at the Staff Academies, reading, an activity which was seriously neglected before, suddenly becomes important for the officers. Some of them give it up again after they have qualified, but it must be said that staff officers are still the best-read group in the Turkish Army.

In any case, the staff officer cannot immediately set his books aside even if he wanted to. According to a recent regulation, he has to sit an examination every time he is promoted. Moreover, he is expected to attend a five-month refresher course at the Armed Forces Academy within the ten years following his qualification. Seventy-five staff officers participate in these courses, which are held twice a year. Emphasis is given to such subjects as air–land–sea tactics, joint operations, campaign planning, military evaluation of neighbouring countries, military law, areas of global conflict, strategy, new concepts and doctrines, and security matters. Most of the participants are from the ranks of major and lieutenant-colonel and are accommodated in the guesthouse.

THE ACADEMY OF NATIONAL SECURITY

This is yet another unit within the complex of Academies, which convenes once a year for a five months course with the participation of twenty civilian officials and ten high-ranking officers (mostly colonels and, occasionally, one or two generals and admirals). This is intended as the highest

---

* An officer may sit admission tests once a year for six years running. To be eligible for them he must first have obtained 70 out of the annual 200-points of the credit system and bring a recommendation from his superior officer, and he must not have been disciplined or confined to barracks for more than twenty-one days. He may enter admission tests only after these requirements have been met and the approval of the service chiefs obtained.

means for the much needed dialogue between military and civilian bureaucrats, and a channel for the exchange of ideas and information on matters of mutual interest.

> For the Turkish Armed Forces, the Academy of National Security is the highest educational and training institution operating under the command of the Staff Academies. Its purpose is to prepare higher administrative officials for joint planning in matters of national security, and for coordinated work in the implementation of plans. It also aims to establish a common understanding between the state institutions and bodies with respect to national security and defence, and under given directives, and in coordination with the National Security Council to contribute to the solution of problems relating to national security.

The Academy is the only military educational establishment in which civilians take part, and officers have the chance to work with them for a lengthy period. The majority of the civilians invited to attend are ambassadors, under-secretaries of Ministries, their deputies, general directors, provincial governors, sub-governors, regional directors, heads of government departments.

During a tour of the Academy, officials stated that their real purpose was to train top administrators in an awareness of the principles of Ataturk. They also added:

> We study international trends in politics, economics and military matters. We examine the situation in neighbouring countries and discuss home affairs. Our discussions, more in the form of seminars, cover national policies, aims and interests and are kept confidential.

To date 400 officials have attended the Academy and some have kept up friendly relations even after graduation.

> We tell civilians about our world as they tell us about theirs. We try to work out ways of overcoming the difficulties facing Turkey. We all love our country . . . It's a pity that the civilians generally

send us officials who are not on active service. We would get closer to our goal if that policy changed.

DIFFERENCES FROM WESTERN STAFF ACADEMIES

The Staff Academies in the West differ from their Turkish counterparts in many ways. First, training is not so intensive and lasts one year instead of two. The second point concerns the selection of officers. Those who are admitted to the Turkish Staff Academies are first lieutenants and captains aged about 30, with an average of six years of regimental service. In the United States, future staff officers are selected from majors, whose average age is 35–7. Age and regimental experience account for greater receptiveness during the courses.

In the West (for example, in France) the preliminary selection is made by the commander of the regiment, who decides who may qualify. Success in the examinations and a good record are not the only criteria for candidacy. Officers must receive an appointment before they can sit the examinations, which last a week in France and four days in Germany and Britain. More importantly, there is also an interview to test many other aspects of the officer, from his intelligence and wit to mathematical aptitude.

Differences evident at the beginners' level are due to the fact that Western officers have a higher standard of general culture. Being fully exposed to world problems in the Western media they are much better informed and more cultivated than their Turkish counterparts, who are obviously less privileged in these respects.

Concerning the differences in teaching methods, a Turkish general who had qualified in the French Military Academy made the following remarks:

> They like to stimulate debate among the officers, who are urged to think for themselves and discuss their views. Our system, on the other hand, favours lectures, offers given topics, a certain orientation, and proposes a method of solution. We put more emphasis on the solution than on pressing for alternatives. They refer to 'one method of solution', unlike our 'method of solution

according to the Academy'. They are not given ready-made solutions, but like to be as flexible as possible.

The way of handling topics is also different:

A situation is proposed: e.g. 'the Soviets have penetrated Anatolia via Bulgaria and Syria'. Every officer devises a way of countering the situation and explains it to the class. Each is put to a vote, and the officer whose proposition wins the greatest approval is given the chance to set up his headquarters, to choose his men and develop his tactics. The purpose is to encourage officers to use their initiative. But for us, the important thing is to fill the gaps through intensive teaching.

The Western Staff Academies give a hearing to all political points of view, particularly liberal ones. In Turkey, however, the Staff Academies remain as closed to political views as are the Military Academies. This attitude, which is said to be the result of conditions peculiar to Turkey, allows for moderately conservative opinions to predominate among the lecturers.

Another distinctive aspect of Western Staff Academies, which concerns the economic power of the countries involved, is that practical application is as important as theory. Much time is spent on field-trips and manoeuvres, whereas in Turkey, teaching generally remains theoretical.

The unique feature of the Turkish Staff Academies is that they have combined all three forces on the same campus. Each finds the opportunity to be informed on the means and resources of the others, and army, navy and air force officers become more closely acquainted. Another strength is that the general educational standard of the Turkish Staff Academies is higher than some of its counterparts in Europe, and much closer to that in the United States.

As I was leaving the campus of the Staff Academies, I could not help wondering why this institution was not better known by the public. At the gates, I came across an officer who was exceptionally relaxed and smiled as he walked. He greeted me and said he had just found out that he had qualified as staff officer. His eyes shone with satisfaction. I asked him how it felt.

'Great. It's very exciting to think that the way is now open for me to be a general. I'm a staff officer now.'

I was not quite sure at the time what he meant, but I wished him luck all the same.

Would he be able to make full use of his qualification and enjoy it, or would he be lost in paperwork?

13
The Gendarmerie: An Unfortunate Service

If an opinion poll were held on the gendarmerie in Turkey, the results would not be very favourable. Their image with the urban population, who do not often meet them, would not be so bad, but would still be described as 'grim' or 'severe'. Civilians regard the gendarmerie as a force they want nothing to do with. In areas under martial law, some are known to have given themselves up to officers rather than to the gendarmes. In the rural areas people are truly frightened by the gendarmes. Many prefer to be dealt with by the police. The general opinion is that in the villages the gendarme wields more power and is more in control than the army officer.

A brief study, however, shows how the reputation of a service can suffer as a result of being misdirected. It also easier to understand why they have become notorious for 'ill-treatment'. The gendarmes may be said to be atoning for the mistaken policies of the past and present, so it is appropriate to describe them as an 'unfortunate service'.

As a profession this force suffers many contradictions. Although the gendarmerie is part of the Armed Forces and ranks fourth in the general hierarchy, it is considered more of a police force than a military one by the army. Its function is seen as 'assistance in maintaining discipline', but this does not bring it any closer to the police, who generally distance themselves from the gendarmes.

In many ways the gendarme is more realistic than the officer. He is more involved in civilian affairs and with laws and regulations. The officer tends to solve matters by carrying out orders, whereas the gendarme has a much wider perspective on events and must be much more careful about the law.

The position of the gendarmerie is unique in Turkish bureaucracy, in that its duties and the orders it receives derive from a variety of sources. It is attached to General Staff Headquarters in respect of training and special duties, to the Land Forces in regard to weaponry and equipment, and to the Ministry of Internal Affairs in its duties concerning public security and order.

Its duties may be summed up as follows:

Administrative: to ensure security within the area of duty; to track down smugglers and hand them over to the judicial authorities; to provide external security for prisons.
Judicial: to track down, apprehend, and investigate law breakers.
Military: to track down fugitive conscripts; to secure military zones; to enforce military prohibitions; to collect conscripts; to secure borders and prevent smuggling; to serve in the commando units and the air force. Military duties are assigned by the General Staff, and others by garrison commanders, recruiting officers, provincial administrators, and state prosecutors.

The gendarmerie operate in about one-third of Turkish territory: they are responsible for public security and order in all areas that lie outside the jurisdiction of the police, that is, particularly in the rural areas. They have headquarters in Istanbul, Izmir, Konya, Adana, Ankara, Erzurum, Kayseri, and Diyarbakir, and 3,600 posts (excluding those on the borders) scattered all over the country. For instance, the official despatch of a law recently passed in the National Assembly to a distant mountain village is also among the many duties of the gendarmes.

The force is 115,000 strong but with only 1,300 officers, 7,000 NCOs, and 1,000 reserve officers who serve as technicians. The root of the problems concerning the service lies in the fact that the main burden falls on the 105,000 who are conscripted for 18 months.

After a preliminary training, we have no choice but to arm the conscripts and send them off on duty. As they know nothing about laws or regulations their manner of enforcing them is rough. Hence, the negative image.

The concscripts generally perform their duties under the command of NCOs, but as there are not many of them, the conscripts sometimes have to track down law-breakers on their own.

As a result of inadequate training the NCOs set a poor example to the conscripts serving under them and used to treat them harshly. But the situation has taken a turn for the better. Improvements in training will resolve difficulties, though it is still not so easy to wipe out unpleasant memories.

Unfavourable impressions of the gendarmerie date back to the nineteenth century. The force was founded on 3 November 1839 and was reorganized on the basis of the French Gendarmerie in 1879. It was called on to perform the heaviest duties, such as serving in civil conflicts and uprisings, in the collection of taxes and the pursuit of criminals. It became the agent of brute force for the central government.

However, the most difficult period began with the Republic, during the early days of which the gendarmerie was the only force to ensure public security. In the face of widespread unrest throughout the country, the gendarmes had to act as a police force, intervening in all kinds of events which endangered internal security. Emergency situations arising from reactionary activism and political opposition had to be dealt with severely and without much regard for the rule of law. The events at Dersim in 1938 are still within living memory. The unfortunate legacy of this period is due to the role that the gendarmerie was made to assume in enforcing the principles of Ataturk, in suppressing all forms of opposition, and in using force to put down uprisings.

The duties of the gendarmerie remained equally harsh even after the Republic was well established. Under the government of the Republican People's Party, the gendarmes were called upon to collect compulsory levies. Encumbered with heavy duties and under the leadership of poorly trained officers and NCOs with no regard for law or human rights, they came to be associated with deep-seated public fear.

Today the position of the gendarmerie has undergone a complete change. Future gendarme officers are selected from the cadets in their second year at the Army Academy, the selection being based on their aptitude for law. Subsequent training is outlined as follows:

> After the Army Academy, officers first attend foundation courses at the Infantry School for eleven months, and are then trained for three and a half months at the School for Gendarmerie, where they receive instruction in laws and regulations. This is followed by practical training in apprehending criminals, carrying out searches, and administering stations of the gendarmerie. The officer receives his commission after having attended a four-months commando course.

Personnel regulations regarding such matters as promotion and accommodation for the gendarmerie are the same as those for the General Staff. The only difference is that gendarme officers receive a monthly compensation of 6,000 TL from the General Directorate of Security for working overtime.

Gendarme officers are generally appointed to rural areas and in some cases have to serve five times in the eastern region of the country during their career. There is no possibility of appointments abroad. Those who are assigned to duty on the border posts where there are no garrisons have to work on their own.

The main problem for the gendarmerie today is its inability to keep up with the changes taking place in contemporary Turkish society.

> Turkey has changed radically since the 1920s and 1930s. In the early days of the Republic, for want of a police force, the state had to rely on the gendarmerie. But it so happened that even after the police force was established, the duties of the gendarmes multiplied. The nature of crime, and the ways that criminals operate, have changed in present-day society. Some knowledge of social psychology and criminology is essential, whereas I have

to operate with conscripts who have no such grounding in dealing with crime. How can I achieve anything when my men are far less educated than the people they have to deal with? Some of them even need to be taught Turkish. Is it possible for them to cope with demonstrations at the universities for instance? There is drastic need for a change.

The inevitable conclusion is that the number of gendarme officers must be increased, some of their duties must be taken over by the police, and a system must be developed to carry out operations from central regional stations instead of thousands of gendarme posts scattered all over the country.

We are trying to change our public image. But the only way to achieve this is by bringing about structural changes in the force.

This may be an excellent proposal but also one that requires financial resources which are not available at present.

From my own experience as a resident in Belgium and a frequent traveller in France, I know that the gendarmes in those countries are no more popular than their Turkish counterparts. Their military approach to problems seems to brand them generally as harsh.

The most important difference between the Turkish and the West European (French, Belgian, and Italian) gendarmeries is that, in the latter, duties are carried out by officers and NCOs instead of privates. For instance, the French Gendarmerie consists of 2,546 officers, 76,000 NCOs, and only 9,000 privates, who are assigned mainly to sentry duty and office work. Their contact with the public is kept to a minimum. In the Belgian Gendarmerie, which consists entirely of specially trained officers and NCOs, they have no use for privates. In Italy the gendarme force is in charge of 24 regions and consists of 2,000 officers, 23,000 NCOs, 61,000 expert gendarmes (Carabinieri), and only 24,000 privates.

The advantage in the West is that the police force operates throughout the entire country, that the frontiers are less extensive, and that the stations are closer to one another. The distribution of duties between the various security forces is also more balanced and assignments are received from a central source. Yet another distinction is that the Western gendarmeries are not a paramilitary force but one that provides assistance to the police.

With the changes brought about in 1982, the Turkish Gendarmerie has improved its mobility and is better organized. But it continues to be burdened by too many duties.

14
What Lies Ahead for
the Colonel

For every ambitious colonel from his third year onwards the nagging question is whether he will be promoted general.

> I shall never forget the year I qualified as staff colonel. I had one more year to go before promotion. The Commander of the Land Forces was visiting us at Erzerum, and in the course of an assembly he suddenly addressed me by my first name and spoke to me for five minutes, while other commanders and fellow officers stood waiting. 'I'll be seeing you again', he said as he left. This was the clearest sign to raise my hopes for promotion. From that moment everyone took it for granted that I would be made a general and some even changed their attitudes towards me.

Colonelcy is indeed the last stage for many in the Armed Forces. Those who are not too hopeful of promotion, or are too keen, soon give up trying in anticipation of their retirement. Others, especially staff colonels, count the days to their promotion and do their best to avoid any mishaps that might endanger it. It is an entirely different matter to be a general or 'pasha', as he is commonly called. In Turkish society the 'pasha' is the real soldier, the most powerful and the most influential. He is looked upon in an altogether different light from his subordinates.

There is little difference between a colonel and a general in statutory terms: the relations between a general and his brigade are the same as those between a colonel and his regiment. Nor is there a substantial difference in salary. The real change takes place in the world of the officer who has been made a general. From the moment he receives his stars, he expects to be served instead of serving others. His preoccupations change in nature and in dimension, and this change is prompted not by laws and regulations but by practice. The difference between the ranks of colonel and general involves an incomparable distinction in rights and powers.

The Turkish Armed Forces are highly sensitive in matters of hierarchy. There are bitter memories of captains and lieutenant, colonels in the Committee of National Unity, formed after the military intervention of 27 May 1960, who occasionally had to give orders to their superiors. Great care is taken to prevent a recurrence of this. The significance of a general's rank becomes apparent in the light of the importance placed on hierarchy. The orders of a general may be contested or controlled only by the few above him. But in order to take his place among the generals who make and implement decisions, the colonel must conform strictly to certain rules regarding his military competence, world outlook, political inclination, and personal life.

From the rank of lieutenant to colonel, promotion is more or less automatic unless there is a grave error or breach of discipline on the part of the officer, but only the colonel who has certain basic qualities may be eligible for promotion to the rank of general. Apart from having achieved professional distinction, he must have lived up to the requirements of an unwritten code which has changed over the years:

(a) He must not be involved in any political activity.
(b) He must be conservative in his social and domestic life: no allowance is made for radical behaviour.
(c) He must be discreet in manner and speech, and in his use of initiative.

In one way, certain qualities which are rewarded in civilian life are frowned upon by the military. Out-of-the-ordinary personalities might prove too risky for the military structure. Furthermore care is taken to avoid problems arising from over competitiveness. There is less tolerance of outspokenness and indiscretion as officers rise in rank. Obedience and conservatism become more important qualities. A lieutenant starting out with liberal views would assume a more conservative attitude as he reached the rank of colonel or general.

The number of generals and admirals at the top of the hierarchical pyramid varies between 280 and 300. It increased as a result of extraordinary circumstances during the political interventions of 1960 and 1971. But because of the shortage of officers in recent years, the promotion of generals and admirals has been slowed down by one year to maintain the balance. From a distance, one might think it was not too difficult to maintain the structure of the pyramid. But who would want to give up his post easily? Moreover, it is difficult for top-ranking generals to decide who among their friends in the lower ranks will have to retire.

The number of generals and admirals has been set by law with a view to facilitating the process. In order to avoid the disarray following the 1960 intervention, which resulted in the compulsory retirement of 255 generals in addition to thousands of officers, efforts are made to keep to the requirements of the structure. July and August are the critical months during which appointments are made. Every year hundreds of colonels who have completed their six-year term anxiously await their promotion, but only 48 go through. Of the many who are unsuccessful, some wait for another year, others retire.

WHO DECIDES AND HOW?

According to the law, the Military Council established in 1972, following the 12 March 1971 military intervention, has the final decision on promotions and appointments of which there are two types. The first are the appointments of service chiefs and army commanders, the second of promotions from lieutenant-general to full general, and from colonel to general.

In practice, the Chief of the General Staff, the undisputed supreme commander known informally as 'the first Chief', decides on the appointments of the service chiefs and promotions to the rank of full general. Regarding the former he consults with the President of the Republic and over the promotion to general he also consults with the service chiefs. But in all cases the final decision rests with the Chief of the General Staff and is therefore uncontested by the Military Council.

A former service chief who was an active member of the Military Council remembers the following points about the meetings he used to attend until a short time ago:

'Any general whose promotion is to be discussed, leaves when the meeting begins. After a brief discussion the proposals of the Chief of the General Staff are accepted. Of course the greatest difficulty is to make a choice between two equals, be they lieutenant-generals or rear-admirals.'

'Don't the Prime Minister and the Minister of Defence interfere at all?'

'No. They make the opening speech then keep quiet. Perhaps they say a few words but they don't express any preferences. If they do, this usually backfires and casts a shadow on the general concerned. Should the Prime Minister have a particular choice or objection, he must reach an agreement with the Chief of the General Staff prior to the meeting.'

'How are the colonels promoted?'

'Each force draws up its own list. In no other Turkish institution is there a more thorough investigation than that on colonels eligible for promotion to general. Their records and qualifications are subjected to a minute and individual study from the rank of lieutenant onwards. Even reasons for divorce are examined. The difficulties here are similar to those related to the promotions of generals: the eligible outnumber the quota of promotions. Decision on the final selection among peers is taken on the basis of additional qualifications such as the knowledge and command of language, etc.'

The status of wives, their education and social background also play a critical role in the final selection.

However, even a meticulous investigation does not outweigh the influence of personal relationships in matters of promotion. A colonel who gets along well with his superiors, who inspires confidence and makes a good name for himself has a greater chance of promotion than another who is equally qualified but less conspicuous.

I don't know the present situation, but until recently, the service chiefs used to prepare a list of their own in addition to that proposed to the Council. We would talk about our own choices beforehand and reach an agreement among ourselves. In those days, the names proposed by the service chiefs would be approved without any objection. As there was no time to dwell on each individual case, the slightest criticism from a Council member could prove disastrous for a colonel with no patronage. The promotions would finally go through by secret ballot. In recent years the participation of all the full generals in the meetings makes it harder for the service chiefs to decide beforehand. Everything happens under the eyes of the 18 members of the Council.

It is extremely important for an ambitious colonel to have good relations with commanders whose opinions will carry weight as the time for promotion draws near. Colonels already appointed to General Staff Headquarters have a further advantage because of their more frequent contact with the highest-ranking generals. Besides a successful record of field service the most important precondition for promotion is to have 'caught the attention' of the commanders.

THE PRACTICE IN THE WEST

In respect of promotion in the lower ranks, the Turkish system is very like those in the West. But there is a noticeable difference as regards the highest ranks and the appointment of service chiefs. In Turkey civilian governments observe strict non-interference, whereas, in the West, Ministers of Defence may accept or turn down the nominations made by General Staff Headquarters. It is also possible for the civilian authorities to take a sudden decision to appoint a new service chief without even informing the Chief of the General Staff.

A second, and more important, difference is that the ranks of colonel and general are not worlds apart. In contrast to the practice in Turkey, it is relatively easy for an officer to reach the rank of major, but much more difficult for him to be promoted colonel. The move from colonel to general does not signify a major rise in authority or responsibility. Moreover, the selection process is over by the time an officer becomes a major and there are fewer hurdles to jump for the colonel who wants to move up. It is quite acceptable in the West to be retired as a colonel because to have reached this rank is a success in itself. As a high-ranking Turkish officer put it, the difference in the Turkish Armed Forces reflects the conditions peculiar to Turkey.

As the time for promotions draws near, colonels who have higher aspirations become obsessive about their careers, devote even their spare time to work at their regiments or headquarters, and become even more detached from the civilian world. Those who are not hopeful of promotion, on the other hand, start preparing for a new life and become more involved in the civilian milieu. But whatever direction he may take, the main motivation of the colonel at this turning point of his career is to do his best to be remembered as a worthy commander.

15
The World of the Generals

I had just been promoted admiral. The first thing I did was to run to my dear old mother and kiss her hand. She looked at me attentively for a while, and then she sighed and said, 'That's all very well, my son, but you had better tell me when you'll become a Pasha.' Basically she was right. To become an admiral does not really count in our society. The important thing is to be a Pasha.

The status of Pasha is an institution inherited from the Ottomans, but in that period not only the military but also high-level bureaucrats were addressed as Pasha; thus no terminological distinction was made between civilian and military. The overall authority was the Sultan, who was both commander-in-chief and ruler. Although the present situation is different, the status of Pasha is still thought to be the highest. Once officers in the Republican army are promoted to the rank of general ('Pasha') their overall situation and environment as well as their authority undergo changes far more radical than those experienced by the other ranks. Of the 800,000 members of the Turkish Army, approximately 70,000 are commissioned and non-commissioned officers, whereas the number of generals at the top of the pyramid is only 285.

As mentioned in the preceding chapter, in statutory terms there is no great difference between being promoted to colonel or general. For instance, a general's salary does not rise several times over. His influence, however, does increase with his function.

When I became a general, what impressed me most was finding myself suddenly in command of a brigade 6,000–7,000 strong, and accordingly of thousands of armoured vehicles, tanks and weapons. I also had the power to authorize transactions involving hundreds of millions of lira. Especially in deep Anatolia this office makes you the second most important person after the governor. Protocol changes completely. The higher the rank the greater the duty, authority and responsibility, by a multiple increase. The office of commander of a division of 10,000–15,000 and a corps of 15,000–70,000 brings with it an authority and responsibility similar to that of governing a fair-sized state.

I was a corps-commander in Erzerum. People didn't know me by my name: they called me 'The Governor of Erzerum'. According to them I was number one protector of the state. To a general the doors are open to the highest duties in the General Staff and the Headquarters of the Service Commands. One gets a little closer to commanders who make the decisions. When you are in field service you advance in protocol, and important people compete to meet you: similarly when you go to Ankara and take up office in the General Staff or the Headquarters of the Service Commands you begin to move in different circles. You gain the opportunity to dine with a service chief or the Chief of the General Staff, which was hardly possible when you were a colonel, and your opinions are sought more often. It may be that decision-making on the policies of the Armed Forces lies not with you but with someone else. Nevertheless you still retain your influence and status.

He who until recently was serving his commander is now being served. His headquarters are more capacious, he no longer queues for accommodation, and whenever possible he is allocated more spacious quarters.

Every general is furnished with an official car (a Renault) and a chauffeur. He had been given official cars in the past depending on his duties, but the general's car with several stars on its number-plate is quite another thing. He is also given an adjutant, and the number of his orderlies increases. He no longer has to go to the bank to deposit or draw out money, nor does he have to chase up theatre tickets, or queue for football matches. Everything is done for him. Phrases such as 'Sorry, no room!' are unheard of in the officers' clubs and recreation centres. Even in dining-halls where there is only one waiter to every ten people you have one to yourself, and he will even toast your bread if you wish. All these may seem trivial details, but they give colour to a man's life. They are the most important indications of his rise in status. Who wouldn't be pleased?

When the officer becomes a general, and particularly at the time of his promotion, it is the civilians who pay him most attention.

> All at once I found myself surrounded. Distant relatives whom I'd never met before and old class-mates began to emerge out of the blue. After I became a general I started to receive flowers and congratulatory telegrams and phonecalls at my every new promotion or appointment. I was surrounded by a 'halo of attention and affection' from all sections of society. My officer friends also showed interest and congratulated me, but the civilians surpassed them all. I would never have imagined that they would pay so much attention to those who become Pashas. I soon figured out the reason.
>
> What you observe in the attitude of civilians to generals is an unprecedentedly close relationship which an officer has never before experienced in all the thirty years of his career. People from all parts of society – the private sector, the bureaucracy, and those from his former neighbourhood – all have a 'a friend who is a general'. They like to keep in close contact with him all the time. They compete with one other to be able to say, 'The other day I spoke to Ahmet Pasha'. . . ., or 'I'll tell Ahmet Pasha about this', or to use his name (usually without his knowledge) to solve their problems that are caught up in the wheels of the bureaucracy, and to ensure their future positions.

'My dear sir, what a wonderful idea! We do need people like you.'

'My dear sir, please don't put yourself to any trouble. I'll send my yacht to fetch you. I want to benefit from your opinions.'

'I'll be very disappointed if you don't come to my party.'

After a while, a Pasha becomes a person whose 24 hours are fully occupied, especially if he has been appointed to a city. Presents, invitations and compliments pour in. Some generals, not being accustomed to this kind of life, do not much care for it. But others do, and make the most of their positions in society and in the official protocol.

Certain sections of civilian society would do almost anything to have a general on their side, converse with him, or at least be seen with him.

> When I was commander of the Northern Maritime Zone, I had no idea that protocol required so many duties from me and I found myself running from one invitation to another. The wheel gets hold of you and grinds you down till in the end you come to your senses. You are only human and you can't bolt your doors like monks who have renounced the world. The important thing is to strike the right balance.

Most of those who are or have been generals or admirals agree that the excessive interest of civilians arises from their desire to be close to the man with 'authority and power'. The general or admiral suddenly has many more contacts with civilians and government staff because of his prominent position in protocol and his involvement in affairs of state. From now on, at receptions which are for him compulsory, he is pushed into corners and questioned about current government policies or the way the country is going. And because he cannot avoid such queries he becomes more involved whether he likes it or not: it can even be said that to a certain extent he becomes politicized.

When an officer becomes a general with a new rank every four years, his wife's rank also rises. For some this makes no difference, but others find it upsetting. However, for some it means an entry to the modern world, which they welcome.

> According to our customs, the respect shown to a high-ranking officer is shown equally to his wife. My husband was a brigadier. I no longer had to wait my turn at the hairdressers. I could even summon him to my house if necessary. Daily invitations to tea or dinner increased immediately. Seating arrangements also change according to your husband's rank. You are given the seat of honour. But if he is only a brigadier you may not precede the Lieutenant-General's wife. At first I could not understand what my husband meant when he said, 'Amazing! More people than

ever are saluting me.' Later I began to understand when I received the same attention.

To be close to a commander's wife, to have contact with her, or even to be a friend, is a social requirement and also establishes a system of mutual support that may be needed in time of trouble.

It is customary to show respect for our elders at home and elsewhere, and we take great care not to upset this system. Our husbands may be a bit flexible about distinctions in rank, but we are more rigid and are very careful to maintain this system. When you eventually rise to the same grade you begin to expect the same respect you have shown to others in the past.

Life outside the army is also rewarding for some wives. You enjoy the photographs and articles published about you in the society columns of the press, and being the president of the administrative council of this or that charitable foundation. You also like to be seen with celebrated performers who address you by name and announce that they will sing a song or whatever for you at parties in famous places you haven't visited before, because you couldn't afford to.

The commander's wife, who has withstood all the difficulties alongside her husband and arrived at this rank, does not want his office to come to an end. The most important motivation for him is to aim for a further step up the ladder, not to leave such a colourful world in four years. Since the pyramid gets narrower near the top, the competition for promotion increases. From now on, even the smallest errors are unacceptable and great pains are taken to avoid any move that would appear to be opposed to the conservatism of the top brass.

Along with its pleasant aspects, the career of the general has additional difficulties for some.

The rank of general increases the pressure exerted by your uniform which imposes the responsibility of paying attention to your behaviour throughout your whole career. You can't go wherever you like. For example, once I had to go to Antalya. I didn't know what to do since I couldn't take my official car. I couldn't have taken a bus because eminent generals never take buses. I had to give up shopping. You can hardly go to the cinema or the theatre.

And as the general is promoted the responsibilities grow. A general is obliged to obey more rules. For example, no matter how eminent he is, if he gets sick, his chances of promotion immediately go down according to the nature of his illness. A general has to be always sound, healthy and obedient in all respects, and a good administrator.

Moreover, during this period of satisfaction and difficulties that the career brings with it, the dark clouds of retirement also begin to gather.

The majority of generals prefer to take field duty rather than function as one of the General Staff, although the latter is more prestigious. One has power in the field but its moral requirements are more intense.

I was going to be appointed either to the Army Corps or to the General Staff and would have become one of the chiefs of J. Being the chief of J is also important but always and in all nations the Army Corps is more onerous. When they told me that I had been assigned to J-5, I was quite disappointed . . . But we never dispute our assignments, and only try to do our best.

This commander was going to take up duty in the General Staff. He soon realized that the world of the General Staff was very different.

THE GENERAL STAFF

The various Headquarters of the Armed Forces are located in a cluster of dignified 3–4-storey public buildings on the way to the government departments district in Ankara. With sculptured eagles at its gates the most impressive of these buildings is the Air Force Headquarters, while the least ostentatious belongs to the General Staff Headquarters – the 'brains' of the second largest army in NATO and the seat of the supreme 'Commander'. If the rank of full general is the highest a commander can reach, the highest position he can occupy is that of Chief of the General Staff, the ultimate stage in the career of our hypothetical 'Commander'.

I was taken to the Chief's office up a flight of stairs, evidently used exclusively by visitors, and ushered through a door guarded by two tall sentries into a waiting room where I was offered refreshments by the adjutant. I was excited at the prospect of meeting the 'Commander' of the highest rank and position. I had been following him like a shadow ever since he was admitted to the Military School, had shared his initial loneliness there, his excitement when he first put on his uniform, his gradual adjustment, and his highly disciplined life in the Military Academy. Now we had almost reached the end of the road – and the end of this book. Although I was an old acquaintance, he was not likely to recognize me. But I would first watch him quietly, then talk about other matters and take my leave.

The adjutant walked in: 'The Chief of the General Staff will now see you for 15 minutes. Unless he indicates otherwise, please take care not to stay longer. He has other appointments.'

I was somewhat surprised by the simplicity of the average-size office where the Chief's desk stood on the right with seats on the left. The most interesting feature was the telephones of various colours providing direct lines to the service chiefs and army commanders. We shook hands. Instead of the nervous cadet of 40 years ago, I was looking at a solemn commander with a deeply lined face. He asked me politely what I had to say.

THE 'FIRST CHIEF'

The Chief of the General Staff is usually referred to at Head-quarters as the 'First Chief', and his deputy as the 'Second Chief'. The First Chief commands the constant attention and respect of all ranks in the 800,000-strong Turkish Armed Forces. As he can mobilize the forces by a single order, he is the unconditional Commander-in-Chief who has the final say in all military matters. His authority is undisputable.

The status of the Turkish Chief of the General Staff is different from that of his counterparts in other armies in that he is directly responsible to the Prime Minister instead of to the Minister of

Defence. He immediately follows the Prime Minister in state protocol which assigns the first four places to the President of the Republic, the Head of the Constitutional Court, and the Speaker of the National Assembly and the Prime Minister. On account of particular conditions in Turkey, the movements and speeches of the Chief of the General Staff command the attention not only of the Armed Forces but of all sections of Turkish society. From a certain angle he may be considered the 'most important person' in the country.

His responsibilities are in no way less extensive than his power and authority. Events such as the death of a conscript in the southeastern provinces or the sinking of a small boat during naval operations, and the military modernization programme that costs billions concern him personally and become part of his daily life in the course of his four years of office.

The personality of the Chief of the General Staff is also of great importance as it has a bearing on Turkish politics as much as on the policies implemented in the army. If, for instance, a Chief of the General Staff is particularly concerned with language education, this will be reflected in an increase in the hours of language teaching in the Military Schools. If another shows interest in armaments or construction, then funds are spent for those purposes rather than for others. Moreover, the policies he decides on or directs will be affected by the harshness or flexibility of his temperament, which will also influence relations with the civilian government and the Prime Minister.

According to the law the duties of the Chief of the General Staff are defined as follows:

> The Chief of the General Staff determines the principles, prio-rities, and main programmes related to personnel, intelligence, operations, organization, education/training, and logistics ser-vices in preparing the Armed Forces for war.

In other words, he is responsible for training, armaments, and all forms of preparation in times of peace. In war, according to the 1982 Constitution, he takes over the Supreme Command from the President of the Republic.

The power of the General Staff is best explained by a full general who, for many years, occupied a prominent position:

The power of the General Staff lies in its hold on the budget. Resources are not in the hands of the Ministry of Defence but of the General Staff. In addition, the responsibility of planning for war and of assuming supreme command lies with the General Staff. Counterparts in the United States and Europe are not so powerful, because, unlike our system, civilians there take a more active role in decision-making.

Indeed, the practice in many other countries, is exactly the opposite of that in Turkey. The office of Chief of the General Staff is subject to rotation among the armed forces every two years. The view that an officer who has been appointed commander-in-chief of the navy or the air force may not be sufficiently competent to take on the duties of Chief of the General Staff is not favoured.

'If he were not competent enough, he would not have been appointed commander-in-chief in the first place. If he is sufficiently qualified to be commander-in-chief, then he must be equally well qualified for the office of Chief of the General Staff', commented an officer at the Pentagon. He went on to point out the importance of rotation in securing a balance between the services:

In the USA, as everywhere else in the world, there is great rivalry among the three services for their share of the budget. Furthermore, each is convinced of its own superiority. As under our system threat evaluations and the development of defence systems are the responsibility of the civilian authorities, the rotation of Chief of the General Staff among the three services allows each to have a say and reduces friction. It is true that in the end it is the smallest service that gets the biggest share, but at least the others don't feel excluded.

The Turkish Chief of the General Staff is chosen on the basis of seniority: even one day matters in deciding who is to be appointed to the office. Certain practices which in the past favoured the precedence of army officers over their air force and naval counterparts in promotions have now been eliminated to ensure equal opportunity for all three forces. Although in practice the Chief of the General Staff is still appointed from the army, changes are predicted for the late 1990s. These depend on whether a more balanced view of the services will replace the attitude that rates the competence of the army above that of the navy or the air force.

HEADQUARTERS AND J-CHIEFS

The J-Chiefs (J from 'Joint', which is a term taken over from the American system) constitute the foundations of General Staff Headquarters. Their responsibilities are outlined as follows:

J-1 lays down the principles related to personnel, by which the headquarters of the service commands proceed with appointments and promotions.
J-2 collects internal and foreign intelligence, and evaluates reports from military attachés abroad. This is one of the most important offices in the Armed Forces because it provides the basic information underlying 'threat evaluations'.
J-3 is in charge of operations, that is, training and organizational policies. Its most important function is to plan for operations in war, and military manoeuvres.
J-4 is in charge of logistics and all matters concerning the requirements of the Armed Forces, from weapons and ammunition to food.
J-5, the most important office, works out strategic-military policies in cooperation with the Ministry of Foreign Affairs, and is in charge of threat evaluations and Strategic Target Planning. J-5 is also responsible for budget allocations and military agreements.
J-6 and J-7 are in charge of communications and electronics and of studies in military history and strategy, respectively.

The Turkish Military Representative in NATO, in Brussels, and the National Military Representative in Supreme Headquarters Allied Powers Europe (SHAPE) are both attached to the office of the deputy Chief of the General Staff.

Duties are assigned by the J-Offices to the relevant department heads, either on their own initiative or at the request of the office of a service chief. The departments consult with the relevant armed service or all three services and draft their proposals, which are subject to discussion before being submitted to the J-Chief. If there is a consensus on the subject, he merely revises the draft and sends it on to the deputy Chief of the General Staff. If there are differences of opinion, the matter is discussed by the deputy Chief, the relevant J-Chief, and representatives from the service commands. Finally, the draft is submitted for the approval of the Chief of the General Staff. If approved, orders are given for its implementation; if not, it is returned to the J-Offices for change or revision. In the event of a persistent objection from any one of the forces, a service chief can try to persuade the Chief of the General Staff to change his mind. Should this be impossible, there is no choice but to accept his final decision.

The prime concern of the General Staff is to ensure that the mechanisms of the Turkish Armed Forces operate without hindrance. Occasional problems are encountered in conveying instructions from the higher ranks to the rank and file, but in general, the Armed Forces can be considered the most efficient institution in Turkey.

Orders signed by the Chief or deputy Chief of the General Staff are communicated downwards without delay. They have developed a marvellous method of classification which is simple and clear. Sometimes notices which have to be read out to the conscripts or instructions for NCOs are filed away and forgotten but these concern less urgent matters.

The efficiency of the communications system, which is an important contributory factor to the power of the Armed Forces, remains unique in Turkish bureaucracy, even including the office of the Prime Minister.

16
Retirement: Death of a Soldier

I did not recognize the Commander. We had lost touch with each other for quite a long time till one day I had news of him from a common friend. 'He has retired and never goes out. He's just waiting to die.'

In one of the more out of the way quarters of Istanbul, I called on an apartment which overlooked the balconies of another block of flats. When the door opened I could not believe my eyes. Where was the lieutenant, that figure of authority? Where was the resplendent corps commander, the general who with one word of command had got thousands to their feet, who had been so respectfully greeted in the street? Now, no longer the man who had held the fate of Turkey in his hands at General Staff meetings, one word from whom had made political parties and assemblies tremble!

I looked around; there was no longer an adjutant standing bolt upright at the door, awaiting orders, nor a car and a chauffeur for the Commander, nor a bodyguard of soldiers in their jeep, armed and ready to protect him. He smiled, guessing my thoughts.

'They've all gone, everything's gone – vanished overnight. I was left here, quite alone. It's not often anyone calls on me or inquires after my well-being. There's hardly anyone left.'

I went in: it was a modest, medium-sized flat. His uniform was still hanging up. 'It's too upsetting. I should take it down.'

He found himself unable to go out. 'I feel everyone will be sorry for me. Besides, how can a man who has given orders to huge army corps and armies go and sit in a coffeehouse? In the past when I entered they would rise to their feet but now they just say something like "Never mind"! It took me a year to

189

find out how to draw money from the bank and where and how to have a shave. It's particularly difficult to climb into a bus with all that pushing and shoving – it gets me down . . . So I go out as little as possible.'

His eyes grew wet. I had never seen the Commander in this state.

When retirement arrives for the soldier, some deal with it easily, others find it hard. It is a disturbing phase in their lives and very difficult for them to come down to the level of civilians and mix with them. In all their years of training they had regarded themselves as superior.

And the difficulty increases in proportion to their rank at the time of retirement. When they have reached the rank of colonel, retirement opens the way to and ensures an easier adaptation to civilian life. A fairly lengthy period lies before them in which work can be found without too much trouble and dialogue can soon be developed. Those who leave the army when they are NCOs or lieutenant-colonels adapt even sooner if they are expert technicians. But it is more traumatic for the higher ranks.

The Commander sat down opposite me.

'Do you manage financially?'

'No. I was lucky enough to have bought this place for myself, otherwise I would have been destitute. My forty years in the Army brought me a retirement pay of approximately 170,000 TL; there's no great difference between what I have now and when I was in service. The difference lies in the perks. No car, no adjutant to do all the jobs, no orderly, no house in which I could live on 10,000 TL a month. Civilian life brings expenses I didn't have in the past. And they keep going up. What's most serious – my comrades are no longer around, I find myself quite alone. Of course this is normal but I wasn't prepared for it and don't take to it easily.'

'Has no-one looked you up?'

'In the early days they kept coming around and saying how sorry they were and that someone like me should not be dispensed with, and I had many telephone calls to say what a shame it was I had been dropped. How can you retire someone because there are no more available positions? What a lot of money this country lavished on me, and then chucked me away.'

'Don't your army friends visit you?'

'Thank Heavens, they do, but they all have their own problems. Besides, they meet together like a retirement club. Everyone prefers his own age group. They play cards and chat. How can I play cards with people who worked under me?'

'Don't you benefit from army privileges?'

'I've right of entry to everywhere, the officers' club, the army shop, holiday camps. My friends go there, but I can't. I should have to queue up with others, or be asked to "sit here, not there". It's natural, of course; when the uniform goes this business is done with too. They always behave with respect and address me as "my Commander", but there isn't the same awareness in their eyes as when I was their actual commander. I'd rather not go there any more.'

'But there's no need for such a trauma . . .'

'Perhaps it seems easy to you. But I'm only human. Having left behind an enormous army with all its authority and responsibilities, and even its showy displays, I'm not yet used to an ordinary man's condition. I wonder when I will be! I can take offence at anything and any word and regard myself as a useless human being with nothing to offer society.'

'Would you like to find work?'

'There was no problem about that. It turned up as soon as I retired. The private sector was really keen. How many times they offered me membership on a board of directors! Other friends accepted without hesitation, but I could not. Once you've been a commander you can't take on any old thing. Society imprisons you, and as well as those who ask "is this work worthy of a great commander?", other retired army men put pressure on you, commenting that it's shameful and beneath you. There'd be no problem if I'd been of lower rank.'

Retiring soldiers are not so very different from other social groups, of course. In the first years the retirement pension is fairly satisfactory but in time with inflation it begins to shrink to an inadequate amount. The private sector has taken retired generals on to their executive boards to facilitate their dealings with the government, and not entirely without success. Those who suffer most are the lower-ranking generals. When they reach a stage where they cannot live on their pensions and are disinclined to contact the private sector, they soon face a dilemma. It is extremely hard for a lieutenant-general who has been in command of an army corps, or an admiral retired from a naval command, to work in the local authority's archives or in a bank for supplementary income.

> It's impossible. I can't demean myself like that when I've been in charge of thousands of men. And if the rank has been high, some firms don't find it easy to offer employment. So we're forced into total isolation.

The soldier who all his life has had little contact with civilian society pays the price for his exclusion at the time of his retirement. The Western soldier is more adaptable. He regards soldiering as a profession whereas the Turks see it as 'a way of life', which, if honour and acclaim and the uniform are removed, inevitably ends in a state of shock. Turkey trains its officer to be above and outside society; it constantly exalts him to the skies, then one day it suddenly abandons him to a lonely life in a completely unfamiliar environment. If he were made to face realities earlier on, he would suffer less in retirement.

> It took time to adapt to a disorganized civilian world I disliked. It's true the material situation of retired civilians is perhaps worse than ours, and, besides, thanks to our contacts, we can find work on the side more easily. What we've lost can't be measured in terms of money; and so, even in our retirement we continue to have contact with fellow retired soldiers.

The conversation went on and on. Before me I saw a troubled man who could not stop telling me that 'this country gave him a grossly exaggerated world, then abandoned him'.

When we parted he said, 'call me up every now and then for a chat'.

One morning, hidden among the advertisements on the last pages of the newspaper *Milliyet*, a little piece of news caught my eye, followed by a row of dots to fill up the space. The Commander had died of a heart-attack.

Next day we gathered in the courtyard of a little mosque in the marketplace at Erenkoy – a coffin draped in the Turkish flag and soldiers standing guard at the head. There was no-one else.

We buried him in silence. He had left behind him only a faded uniform and a modest apartment.

Appendix
How the Machine Works

STRATEGIES/PLANS AND PROGRAMMES

The process by which the Turkish Army began to define its own national strategies and entered the planning and programming stage began in the 1970s. The first few years were spent in coming to grips with the need for planning and programming and from 1980 onwards everything was in hand. The second half of the 1980s, in particular, may be called the period of 'the reconstruction of the Army'. Before that, almost everything had been left in the hands of United States and NATO. Where there is now a national strategy there had been only a NATO strategy. Armies were deployed in accordance with NATO strategy, and instead of national policies concerning the acquisition of arms, Turkey made do with what the Americans provided. Today, however, the Turkish Army can be regarded as a force with national goals and plans.

The Turkish Armed Forces have two separate strategic policies:

(i) NATO strategy. The whole of the Turkish Air Force and a large part of the land forces are committed to NATO tasks. The naval forces participate in the NATO Joint Force (the 'Mediterranean Call Force') which is assembled in the Mediterranean in the event of war.

In peacetime these forces are included in NATO plans and remain under national command but are assigned to NATO if war breaks out. If the Turkish General Staff feels the need to relocate or redeploy its NATO-committed forces, it must report and explain any such move to NATO, though it does not have to obtain prior approval.

The Turkish officer does not identify with NATO. He is incapable of thinking like a German, Frenchman or Dutchman. He is aware that his interests may be at odds with theirs at some point. In contrast, the basic interests of the other NATO countries are identical; officers of other armies are part of the NATO system and they know it. The Turkish officer, on the other hand, sees himself as somebody outside

the system but with an interest in NATO because of common interests
and defence needs.

> Our officers, irrespective of seniority, cannot think in NATO
> terms or in the manner of, say, a Belgian member of NATO.
> For us, NATO is a remote organization that draws up plans and
> takes decisions – an organization we apparently go along with,
> hoping that 'they will be there' when needed. Our thinking is far
> more along national lines.
> The attitudes of our rank and file are no different. I asked a unit
> of 300 men, including university students, what NATO was. Only
> five knew the answer. Some thought it was a sports club. Some
> had never heard of it.

(ii) National strategy. The Cyprus Operation in 1974, and the
imposition of an arms embargo by the US Congress almost immediately
afterwards, exposed the unavoidable need for a 'national strategy'. In
1974 for the first time ever, an outline National Military Strategic
Concept (*Milli Askeri Stratejik Konsept*: MASK) was drawn up. In
the early 1980s this was elaborated, on the lines of MC-299, a NATO
guidance document, and received the approval of the General Staff on
1 June 1986. MASK proceeds from threat assessment to deployment
priorities within the framework of a six-year plan of Force Goals on
weapons acquisition and allocation, to be reviewed annually. The basic
goal is 'the preservation of existing frontiers'.

As noted earlier, an important difference from Western practice
is that the Turkish Armed Forces carry out the national 'Threat
Assessment' entirely on their own, based on intelligence from their
own sources and from the National Intelligence Organization (MIT),
which is always headed by a military man. The civilian sectors – the
Foreign Ministry and the Defence Ministry – take very little part in
drawing up the National Strategic Goals. This may create problems. The
civilian sector takes virtually no interest in military strategy, but in such
a vital matter as assessment of threats to the nation political knowledge
and judgement are as important as military knowledge and experience.
It is a fact that the Turkish Army strives to perform the task of threat
assessment by bringing together all the necessary elements and also to
take political aspects into account. But, despite everything, a soldier
cannot think like a civilian, a politician or a diplomat. This danger is
recognized in Western countries – in the United States in particular –
and their civilian sectors have extremely important roles to play in threat
assessment and even in the determination of the necessary strategies and
arms acquisition policies.

The best known strategists in the West are civilians, often university teaching staff and scientists who make great contributions to the development of strategies. In saying that the determination of strategy in Turkey should not be left to the General Staff alone and that civilians should also be included in the process, we are not advocating that ignorant or incompetent civilians should be involved, but that, with proper encouragement, Turkish universities could produce people with a very good mastery of the subject. It is also possible to set up institutes as in Western countries, to train military officers to learn the civilian method of assessment.

While it is a great mistake for civilians to retire to the sidelines using the argument that 'only the military understand strategy', it is equally mistaken for the military to turn the subject into a taboo on the grounds that 'commenting on strategic topics harms national interests'.

ARMAMENT POLICY

The armament policy of the Turkish Armed Forces dates back to the 1940s. The Truman aid for Turkey and Greece, the participation of the Turkish Armed Forces in the Korean War, and finally Turkey's entry into NATO in 1952 led to increasing supplies of armaments from the United States.

General Pendleton, the Chief of the American Military Aid Mission in Ankara, commented as follows on the situation at the time:

> It could be said that from the late 1940s to the mid-1950s we built up the Turkish Army from scratch. The necessity was clear. But it is also true that we had a very large stock of weapons, ammunition, and other supplies which had accumulated from World War II and the Korean War. We didn't know what to do with them. Some of these were in proper condition, others could be put into use after some minor repairs, so we distributed them to our allies.

As new models replaced the old ones in the US Army, the second-hand arms and supplies were sent off to the allies that needed them, especially to Greece and Turkey.

Officials in the State Department and the Defense Department repeatedly drew the attention of the US Congress to the fact that the Turkish Army could not perform its duties with antiquated weapons, and pleaded for an increase in military aid to Turkey (described as 'the junkyard' by General Haig, the ex-C. in C. of NATO, later Secretary of State).

As American surplus stocks diminished in the 1950s and 1960s, the Americans began to supply arms which had recently been taken out of service in the US Armed Forces.

> Every year we used to submit a list of our needs to the USA. These lists were unnecessarily long, covering everything from helmets to batteries, from heavy ropes to tanks or anti-aircraft. The rule was to ask for as much as possible. The main reason was that we had no armament policy of our own, nor any national objectives, nor even any idea of what we really needed. The Americans shipped over whatever they thought necessary and, regardless of their use, we were only too pleased to be at the receiving end. What's more, everything was donated . . . For instance, the M-48 tanks that were replaced by the M-60 in the US Army were shipped to Turkey. Two thousand Reo trucks and 10,000 jeeps, even if they dated from World War II, were also welcome, We had no trouble at all in finding spare parts for them as they were readily available from stock. The arms we received might be old but they were certainly not obsolete in the USA, where spare parts were still being produced for use in a number of other countries in the world. We had so little planning that we had to be reminded by the Americans which part in the warships or aircraft to replace and when. All the details were recorded in their computers which alerted them when, for instance, replacements had to be made on the F-100s and the warships. Sometimes we would get huge boxes, and we wouldn't know what to do with them until the replacement instructions arrived.

Turkish governments took it for granted that the aid would continue for ever, and those Chiefs of General Staff or service chiefs who did not regard this as 'normal' were usually ignored. The swift changes that took place from 1960 onwards were either overlooked or simply ignored. The United States launched its space programmes, and developed defence technologies that had nothing to do with the old systems. Laser weapons, warships with computerized navigation systems, and tanks manned by two instead of five operators were put into service.

Turkey, in the meantime, was still submitting lists on the assumption that its position in NATO justified them. Military aid was determined by NATO defence plans. Within this framework Turkey, Greece, and Portugal would claim inability to afford the cost of arms, and press for aid. But Turkey's essential needs centred not so much on advanced defence systems against the Soviet Union, received on the assumption that they were 'better than nothing', as on practical ones that would solve ordinary, daily problems. The following comments of a general who served as military representative in NATO in the 1970s are not easy to forget:

> For us it's like having a poor man's breakfast at a table set with gold cutlery. I don't have the helicopters to move my troops, but I do have modern defence systems against the possibility of Soviet aggression. I wouldn't mind that of course if I also had the means to service and repair such weapons. Because of our duty towards NATO we're forced to spend a great deal on these advanced systems, but we can't meet the cost of our essential needs with our already limited resources.

In a top-level meeting in the late 1980s with Richard Perle, the then US Assistant Secretary for Defense, held at the Turkish General Staff to discuss restrictions regarding American aid and the needs of the Turkish Army, the Commander felt he had to make his point clear:

> Don't you think we should also ask ourselves the question whether Turkey got what it really needed or whether it simply had to make do with what was offered?

This is the crucial point underlying the problems in present day Turkish armament policy. However, at the same meeting Perle made his point equally clear:

> One can't say that you have ever been keen to acquire the arms you really needed. One doesn't often come across any one from Turkey at the defence equipment exhibitions in Europe or the USA. You hardly have any contact with arms manufacturers.

In other words, Turkey had relied far too much on American aid.

By the 1970s the defence systems had changed completely. They were far more costly, and, more important, even disused systems were no longer donated. Moreover, spare parts had to be made to order and at very high prices. In the late 1960s the US had adopted the policy of donating money rather than arms. The point in selling disused arms in exchange for the money donated was not only to give with one hand and take back with the other, but to warn that from then on arms would be available only at a price.

From 1972 onwards, however, donations began to be replaced by
FMS ('Foreign Military Sales') loans – a decision prompted partly by
Congress' reaction against donations, and partly by the rise in Turkish
per capita income to over $1,000. FMS loans would cost more than the
average credit on the market, but would provide the warranty of the US
Army and repayments would be on a long-term basis. As loans on the
international market were difficult to find, Turkey decided to make the
best of the FMS system.

The US arms embargo of 5 February 1975 and the suspension of all
military donations and credit came as a shock to the Turkish Armed
Forces but it was also a rude awakening to a state of affairs which was
fast becoming untenable.

Again it was Richard Perle, this time in a statement to Congress in
March 1982, who summed up the situation:

> Almost all the main weapons of the Turkish Army are ineffective.
> These include not only tanks, warships, and aircraft, but commu-
> nication and logistic systems. Most of the tanks are equipped with
> 90mm guns of very limited effectiveness. Operating on gasoline,
> they have short-range mobility . . . Only 1 per cent of anti-tank
> weapons are modern and effective. 89 per cent of short-range
> air defence systems date from the 1940s, and 93 per cent of
> FM wireless equipment is ineffective . . . All the destroyers in
> the Navy are US warships from World War II. 75 per cent of
> submarines are 35 years old and have completed their service
> life. 70 per cent of the fighter aircraft date from the 1960s. In
> addition, military airfields are not equipped with the necessary
> modern defence systems . . . The age and ineffectiveness of its
> weapons not only reduces the fighting power of the Turkish
> Army but greatly increases its expenditure on operations and
> logistic services . . .

There seemed to be 'nothing good' about the Turkish Army except
the heroism and patriotism of the common soldier. This critical juncture
led the Armed Forces to initiate a programme of modernization, in
short, to start planning themselves instead of leaving it to others.

The first steps of the REMO (Reorganization-Modernization) project, to set national defence policy targets and plan a budget, were
taken in the period 1970–5, but full involvement came only after 1981.
On the basis of current prices, the top priorities of the modernization
programme have cost the Turkish Army $15 billion in the period
1985–90, of which about $5 billion have been provided by US and

German aid and the national budget; the rest is to be be covered by future planning periods. Those priorities are:

(a) Fighters for the *air force*. The greater part of US aid and funds from the budget help to finance the F-16 project which aims to achieve a balance of forces in the Aegean by 1994. The second air force priority is projects to improve the radar and command-control systems, and to expand the anti-aircraft system; there is still a shortage of funds for this.
(b) The *Navy's* frigate project, together with the acquisition of missile gunboats, minelayers and minesweepers, and command-control ships with electronic equipment.
(c) Priority projects for the *army* are geared towards replacing M-48 tanks with *Leopards*, replacing the Diesel engines in the existing M-48 tanks and improving their strike force, renewing anti-tank weapons, and modernizing communications systems.

It can be said that Turkey is now paying the price for its previous inertia. If a sound programme had been put into practice in the period 1950–70, its army would have been a much better deterrent and it would have avoided the present huge expenditure on modernization.

> We've given priority to main weapon systems such as warships, aircraft, and tanks, but we can't do much about electronic equipment. We shall fly the F-16 but on land we still have to use M-48 tanks. Our main aim is to meet our own needs and reduce to a minimum our dependence on foreign sources. Instead of purchasing, we would like the technology to be brought here, through joint ventures and cooperation.

The Americans, on the other hand, believe that Turkey is mistaken in aiming at domestic production of its needs:

> Turkey wants to produce everything, not just surplus needs. This is not possible. They will soon realize that it's more profitable to import small weapons or equipment than to invest millions of dollars to produce them in Turkey. No other country has such a policy. For instance, 55 different parts were needed for the upgrading of tanks. When the programme started in 1983 all 55 parts were imported from the United States. At the end of the programme about 30 of these will be produced in Turkey and only 19 parts will be imported . . . Another important point to consider is the need for a strong cooperation between the army and the civilian industrial sector. The involvement of industry will

ensure that the same technology leads to the manufacture of other products. Otherwise expenditure or investments will have been made solely for military purposes.

The modernization process involves a number of other problems, the first of which will affect methods of training for the new electronic systems to be put into service. The most important problem facing the army is the need to adjust to the requirements of the electronic age. Another problem is the rapid change in technology of weapon systems. Before a highly expensive anti-aircraft or anti-tank system can be put into service, a new one comes on to the market.

It would be wrong therefore to expect the modernization programme to be easily achieved. The country has pressing economic needs, but very limited resources. It seems necessary first of all to develop a wide range of ideas and long-term plans concerning realistic threat assessments, and secondly to determine what the needs are and how resources can be utilized. Inevitably there must be better coordination with the private sector and within the Armed Forces to avoid further shortages and the waste of already limited resources. This will be difficult unless modern management techniques are adopted and new methods developed to benefit from the long-term appointment of officers who at present have to return to their units in the field after two years.

The Turkish General Staff are at pains not to waste the taxpayers' money. In Turkey, as elsewhere, there is a good deal of struggle between the Armed Forces and the government, as the High Command press for more funds on the basis of threats to national security and governments are reluctant to jeopardize other economic requirements.

The following statement of a top-ranking general in the General Staff is a precise reflection of the situation:

In Turkey military spending is regarded as current expenditure, and the army as a consumer. As they don't seem to be getting anything in return for the money spent on us, we're thought to be consumers. What should we do? Should we start a war to justify expenditure? We've been given the duty of protecting the nation against threats, so we insist on our demands for weapons and the government always tries to provide us with as little as possible. This kind of struggle seems to be universal, there's nothing odd about it. But in our country it appears to be strange because there's no proper dialogue between the army and the government.

The 1986 budget of the Turkish Armed Forces reached a record level of 1.307 trillion TL. This sum, with the addition of the expenses of the gendarmerie, and aid from the United States (including repayments) and Germany, amounts to more than 25 per cent of the national budget. The modernization programme and the maintenance cost of old weapon systems account for the high percentage.

The navy is the highest spender. Tanks and trucks can be switched off but not ships at berth. Similarly, fighter aircraft have to be flown regularly. But even rifles for the army have to be replaced with more modern types, let alone expensive missiles, whose service life again is limited.

In a country where inflation is high, it is safer to take percentages as points of reference rather than figures, which tend to change rapidly. A general look at the National Defence budget reveals the following percentages: in the period up to 1980, 30 per cent of the budget was allocated to personnel expenses (including salaries, travel funds, and compensations). Since then, due to the growth in the budget, the proportion has fallen, and is now in the range of 20 per cent and likely to remain there for the next ten years.

Purchase of new weapons as part of the modernization programme, spare parts, ammunition, training, infrastructure, and construction works constitute 45 per cent of the budget. In the 1970s the proportion was 34–6 per cent, while in the next few years it is expected to exceed 50 per cent, due to inflation and the devaluation of the Turkish lira. This rate will continue to rise in proportion to the drop in the value of the Turkish lira in the dollar-based expenditures on modernization.

An order issued by the General Staff in 1985 made it clear that priority would be given to the purchase of weapons rather than to construction works. Consequently all construction works, apart from indispensable ones like aircraft hangars, were suspended. There would be no further funding for such ultra-modern buildings as the Headquarters of the Air Force, which cost 6 billion TL. Similarly, with cuts in spending, construction work on housing estates, officers' clubs, and recreation camps has been restricted and standardized.

Budget expenses are regarded as a highly sensitive matter by the General Staff. But has it been possible to develop a system to optimize spending and avoid any waste of resources or to draw up an industrial inventory whereby the Armed Forces can benefit from the private sector? Is the army currently able to make full use of the technology available within its own framework or in the private sector? In the period up to 1980, was it planning and priorities or the negotiating power and persistence of the commanders that were influential in the division of the budget among the services? There are no clear-cut answers to such questions.

THE IMPORTANCE OF US AND GERMAN AID

At present the greatest threat facing the Turkish Armed Forces is the pace of technological progress. Excluding any expenditure on the most advanced weapon systems, the cost of improving fire-power to an acceptable level can go up to $100 billion when all modernization requirements are taken into account. Even the $15 billion up to 1990 is enough to stretch national resources to the limit.

Consequently, US and German military aid is regarded as indispensable. Though insufficient, it is a significant support for the Turkish budget in providing foreign currency for imports.

German aid, which started in 1964, provided genuine assistance in the form of donations. Up to 1986, it had totalled DM 2,050 billion. Although never seen as a large amount, it has provided vital relief in times of economic crisis.

As noted earlier, US aid can be defined in terms of three stages: (i) 1950–64, donations of disused arms and supplies; (ii) from 1965, money donations under the 'Military Assistance Programme' and soft credit; (iii) from 1975, and after the lifting of the embargo, the conditions of aid became more difficult following the introduction of FMS loans.

In the 1980s, under the heavy burden of FMS loans, Turkey (and Greece) began to put pressure on the Reagan Administration to revise its policy. In response, the United re-introduced donations from 1982 and divided the aid package into three: donations, direct concessionary loans (at 3–5 per cent interest), and FMS loans, of which slightly under one-third are donated. Unless the approximately three-to-one ratio in American assistance is changed, it will become increasingly difficult for Turkey to continue with the repayments of FMS loans after 1995. Annual repayments which may exceed $1 billion will heavily pressurize the balance of payments, and the resulting adverse effects on the economy will make it difficult to maintain the level of funding for the Armed Forces.

The most important item on Turkey's agenda therefore is to have the USA transform its assistance into genuine aid. 'Direct loans' at 3–5 per cent interest seemed highly attractive when interest rates on the free market were 15–20 per cent. But at present, although the interest rate is still below the market average, the sheer insufficiency of the loans prevents them from being useful.

The most controversial issue centres on the adequacy of the American assistance for the Turkish Armed Forces. Washington has continued its aid programme so that Turkey can fulfil its commitments to NATO. Over the years, the US Department of Defense has repeatedly argued the necessity of this assistance in view of the country's economic situation and its highly strategic position in NATO defence plans. It has also emphasized the necessity for a modernization programme for the Turkish Armed Forces.

In fact, however, there are no concrete criteria for US aid, which appears to depend vaguely on the relations between Congress and the White House, and also on the capacity of the Turkish government to put pressure on the US Departments of State and Defense. So far there have been limits to the leverage that Turkey can exert by the NATO argument. The following statement by a top-ranking foreign official sums up the rationale often heard in the corridors of NATO Headquarters:

> You're wrong in thinking that Turkey's duty is only to protect us. Your armament is geared not only to NATO defence plans but also to your national policy targets. In fulfilling your obligations to NATO, you're defending yourselves as well. In case of a possible Soviet aggression Turkey will be as much of a target as NATO. You have joined NATO so that you can protect yourselves. Do you think you can absolve yourselves of responsibility simply by leaving the alliance in the belief that you will be spared from attack by the Soviets? We can go no further than providing you with a certain limited amount of assistance. We can no longer provide everything for the Turkish Armed Services as in the 1950s. You have to contribute as well. All we can do is to relieve you of some of the financial burden.

Is it in fact possible for American aid to increase? If so, what would be the conditions for its increase? The most meaningful, and in all likelihood the most correct, response to such inquiries in Washington came from an admiral responsible for strategy:

> If the USA so wished, they would not mind pouring in billions of dollars to your country and fully equipping your forces in the same way as they do in Egypt and Pakistan. The problem is that Turkey is not at the top of our priority list. And the reason is simple. If you were to give us bases in Eastern Anatolia, join the Rapid Deployment Force, provide all the facilities for the 6th Fleet, and allow receivers for Radio Liberty, Radio Free Europe or Voice of America broadcasting anti-Soviet propaganda, then the aid you received would be an eight-digit figure. There's a price for everything, and at present this is as much as you can get for

your support. The assistance would increase in proportion to
to your cooperation. Egypt received billions of dollars by signing
the Camp David Treaty and both Pakistan and Afghanistan are
actively supporting us.

To conclude, in assessing Turkey's requirements for modernization
and the necessary funds, one should not count on foreign aid. The most
important concern for the Turkish General Staff is that aid is unlikely to
continue at its current level, and that it will not be possible to keep up
the modernization programme. However, so long as the aid continues,
no Turkish government will take the question seriously, or think of
long-term plans, or join in a realistic dialogue with the Armed Forces.
The present attitude is to leave everything to the General Staff, and to
avoid the question on the simple assumption that the sum in hand is all
that can be afforded.

Nevertheless the Turkish government tries not to turn down the
demands of the Armed Forces. For example, the 'Defence Industry
Support Fund', set up in 1986, is intended to draw in 250–300 billion
TL per annum from taxes on individuals and corporations, and levies
on alcohol, tobacco, and consumer goods. This fund will help to finance
future projects for the defence industry, and also meet the $10 billion
deficit in the $15 billion modernization programme.

However, the crucial problem lies in the shortage of foreign currency.
The General Staff still recall the 1978–80 period when they had to
press the Treasury very hard even for $40 million to meet the cost of
aircraft fuel. Governments have so far approached the matter with some
reservation, as if they were dealing with requests from the Ministry of
Tourism. As there is no doubt about the current resources and current
military requirements, the only way out of the problem is to reach
a compromise on the basis of assessments of threat, of needs and
resources. It is possible that failure to reach such an agreement will
result in huge financial difficulties which may even necessitate the kind
of concessions that the USA expects from Turkey. The reluctance on
the part of the USA to provide funds seems to predict this.

Talks held in Ankara in February 1986 provide highly significant
clues to this particular problem. At a meeting chaired by President
Kenan Evren, General Necdet Urug, Chief of the General Staff, Prime
Minister Turgut Ozal, and other senior officials were given a briefing
on the modernization programme and its requirements, in the course of
which it was made clear that requests for the funds necessary to purchase
new weapon systems had been turned down by the government. The
Prime Minister spelled out the hard facts:

The priorities of our government are quite obvious. They have to do with the construction of dams and roads, and energy requirements, that is, with economic development. Turkey has many needs but limited resources. We cannot afford to put in anymore.

President Evren turned to the Chief of the General Staff:

You're right. I also had sleepless nights when I was Chief of the General Staff, and was very uneasy when matters of urgency were delayed. But it is quite clear what the available resources are. You have to revise your plans and re-arrange your priorities.

Then he turned to the Prime Minister and made the following unforgettable remark which had possibly never been uttered before:

You must learn to say no. It so happens that first you agree to pay for all that is demanded, but when the time comes for the instalments, you say you don't have enough. That's not the right way. You must figure out what you can afford, and be firm and clear when the demands are beyond your means. Then no-one will try to force the economy beyond its limits.

These remarks are a crucial warning in regard to the future of democracy and to relations between the army and the civilian government.

MILITARY NEGOTIATORS

As pointed out earlier, one of the duties of the General Staff is to sign military agreements and, in particular, to defend its views on military matters in NATO. Thus a Turkish officer may also perform the function of a 'negotiator'.

Turkey's external security is ensured by the cooperation of three institutions: the Ministry of Foreign Affairs, the General Staff, and the MIT (National Intelligence Organization). In the 1980s the involvement of the General Staff increased as political and military complications in the Aegean became top priority. The General Staff is also predominant in the negotiations held every six months with the United States.

The interesting point is that the general Turkish attitude is reflected in the approach of the military negotiators: 'compromise' is generally confused with 'concession', and as a result undue persistence or lack of flexibility can lead to a disadvantageous position. A more flexible attitude requires more training as a negotiator. The following comments demonstrate the restrictions that lead to difficulties:

> Fear of reprimand makes one reluctant to approach the commander for further instructions. The main concern is not to make a mistake. Even if you don't get what you want in the negotiations, nobody will reproach you for abiding by the given instructions. Orders from the top restrict your mobility in the course of discussions.

Military negotiators are generally well-informed and in excellent command of their briefs. But in complex situations, they tend to see matters in black and white, and present them as such.

Compared with other participating officers in open or closed discussions in the NATO framework, Turkish officers (and civilian representatives) tend to feel less at home. This may stem from a national disposition to introversion or from a feeling of isolation from the international framework.

In the words of a high-ranking general, 'NATO is like a computer with a hundred keys. Turkey has confined itself to using only a few keys on this computer. We tend to get lost in detail and never think of getting more out of it.' Indeed, Turkey participates actively in only 25–30 of the 226 NATO committees.

A further difficulty for Turkish officers in key positions is that their short appointments of 2–3 years do not allow them sufficient time to specialize in Brussels or in the General Staff Headquarters.

INTELLIGENCE AND RELATIONS WITH MIT

The Turkish Armed Forces have three main sources of intelligence: (i) each force has its own intelligence officers whose duties are extended, whenever martial law is imposed, to cover 'seditious activities against the state'; (ii) they possess listening services, which use VHF electronic devices with a range of 2000 kms; and (iii) they rely on reports from overseas military attachés. Exchange of intelligence with the United States is carried out when necessary but there are no organic links between the two countries.

The National Intelligence Organization (MIT) serves as yet another source. The relationship between the MIT and the General Staff is not as 'close' as is generally assumed by outsiders. Although the director of the organization has been appointed from active or retired officers since 1960, the predominantly civilian MIT is not allowed to penetrate the Armed Forces, unless given permission. Until recently the National Security Council meetings have been briefed separately by the MIT and the General Staff. While the ostensible reason for appointing an officer as the MIT director is 'to prevent governments from making partisan appointments and political parties from politicizing the Organization', the underlying purpose might well be 'to keep the MIT under control'.

Founded in 1927 as the 'National Security Services', and given its present name in 1964, the MIT is attached to the Prime Minister's office and has two principal statutory duties: to stop any action against the state and to provide the intelligence required by the state. It thus combines the duties of, for example, the CIA and the FBI in the United States and the two similar but separate organizations in France. In its combination of internal and external organizational functions, its only equivalent is the KGB.

The regulations regarding its duties cover just about everything: espionage/counter-espionage in its functions abroad, tracing and uncovering communist, extreme right-wing, separatist (Kurdish), and Armenian groups, etc. internally.

The official who claimed that priorities lay with the interests of the state explained the criteria governing the actions of the MIT:

'We are responsible for the protection of the state. Political parties have no importance for us. Any subversive force against the state and democracy will have to confront us . . . We are not under the command of the military: we cooperate with the General Staff, that's all . . .'

'You are attached to the Prime Minister's office and yet, in all three military interventions, the prime ministers were not notified. Isn't that rather strange? Either you were not informed, and presumably failed to do your duty, or you refrained from notifying them. Which was it?'

'Nobody seemed to take any notice as we reported ten casualties a day in the late 1970s and reminded them repeatedly of what was happening and what to do about it. We knew the date of the coup but we did not report it.'

'But aren't you responsible to the Prime Minister?'

'Our fundamental duty is to protect the state. There was no point in reporting to governments in a state of suspension. We never notified them of any intervention, as this was so obvious that any watchful person would have seen it coming. What more was there to report when a warning memorandum had already been issued by the military?

Let me also tell you that prime ministers don't recognize the importance of intelligence. Apart from Inonu and Ozal, no prime minister ever asked for any substantial service from us.'

'As an organization don't you feel over-burdened?'

'Our duties are very wide-ranging. The lack of intelligence units in the weak police force puts additional pressure on us because we have to take on some of their duties. As everywhere else in the world, our organization is an important and necessary one, provided it is properly equipped and staffed.'

'How many do you have on your staff?'

'I can't tell you.'

'Why do you have staffing problems?'

'Because of inadequate training. Another problem is to disengage ourselves from police work and use our personnel in areas that require better training and qualifications. Otherwise we face the risk of losing our efficiency.'

'What are the shortcomings of Turkish intelligence in general?'

'There is lack of coordination between various bodies such as the Foreign Ministry, the Finance Ministry, and the General Staff. If we were all properly coordinated, our organization would have more time to work on secret information instead of also having to deal with open files.'

'Is your work supervised?'

'Of course it is. All our activities are reported to the Prime Minister's office and to other ministries.'

'Do you cooperate with the CIA or similar organizations on a permanent basis?'

'Like all such organizations we exchange information, that's all. We have no organic links with them.'

One of the main concerns of the MIT at present is to find ways of changing its negative image in Turkish society.

RELATIONS WITH THE MINISTRY OF DEFENCE

We are not in the same subordinate position to a government of politicians as in the USA and Western Europe. We have more autonomy and are more actively involved in matters that concern us . . . We would not allow our Commander-in-Chief to be subordinate to the government. This was established in the constitution after 1960, and there is no question of changing it.

This statement from a top-ranking general reflects precisely the relationship between the Armed Forces and the Ministry of Defence. The principal function of the Ministry is to secure the funds required by the Turkish Armed Forces, and to ensure that sufficient sums are allocated and placed under the command of the General Staff for new defence systems and modernization, as well as for other expenses. The General Staff proceeds to divide up the budget according to its own plans and programmes.

It is interesting to note that, contrary to foreign practice, in NATO meetings the Turkish Chiefs of General Staff have never (since 1960) sat behind their Defence Ministers. This is the clearest sign of the 'difference between their status and duties' as established by the law.

Although the Ministers are responsible for negotiating with their American and German counterparts for further assistance or arms supplies, the Ministry is not involved in decisions concerning foreign aid or its allocation. However, this lack of involvement has forced the General Staff to increase supervision of its own activities.

Much of course depends on the personality of the Minister. Those who prove themselves sincerely interested in the problems of the armed forces, defend their views in the government, and above all show respect, are remembered in a positive light. But both sides have grievances on the issue of maintaining a dialogue between the Armed Forces and the government.

> We, as the General Staff, are in favour of promoting a dialogue. That is why the National Security Council was formed after 1960, followed by the Supreme Military Council in 1972. Their aim was to bring together the Armed Forces and the government to debate the issues. But it didn't turn out to be very effective.

> The military would come to the council meetings fully prepared, and speak with one voice, having cleared up any differences of opinion among themselves. This was not conducive to dialogue.

The fact is that both points of view are justified. The military are too cautious because they cannot quite trust the civilians, while the latter feel too uneasy to be able to convince the other side of their good intentions.

> In the meetings the prime ministers used to talk for hours without offering any solutions to our grievances. We, on the other hand, expected from them concrete responses and a serious working discipline.

A former cabinet minister's response to this is:

The military expected everything to be resolved by a single command in the manner they were used to. But we had to work our way through bureaucracy and the parliamentary system. In government you have to maintain a balance, which is not easy to achieve, or to explain to the military. Besides, they have a healthy centralized system of collecting and assessing all the necessary information. It's extremely difficult to match this kind of preparation within the mechanism of a civilian government . . .